SPEC-RA-SCOPE

THE SYMBOLSIM & SIGNIFICANCE OF SPECTROSCOPY TO SCIENCE & SURVIVAL

By

African Creation Energy

12 - 20 - 2020

WWW.AFRICANCREATIONENERGY.COM

All Rights Reserved
Copyright © 2020, 2021 by African Creation Energy,
Written and Illustrated by African Creation Energy,
www.AfricanCreationEnergy.com

No part of this book may be reproduced or transmitted in any form or by any means, electronic or mechanical, including photocopying, recording, or by any information storage and retrieval system without permission in writing from the author.

ISBN 978-1-716-35203-4

Printed in the United States of America

Sawubona, Sikhona, Shiboka

"I See You,

I am Here to be Seen,

and because You See Me,

I Exist"

~Zulu Greeting

Table of Contents

	Section	Page
1	Pro**spec**tus	5
2	Per**spec**tive	12
3	**Spec**ulation	46
4	**Spec**troscopy	75
5	**Spec**trometry	86
6	**Spec**ifications	91
7	Re**spec**t	103
8	In**spec**tor (About the Author)	113
9	Appendix	125

1. Pro*spec*tus

Behold! Eleven (11) years ago, in the year 2009, I began writing a series of books using the penname and branded term "**African Creation Energy**" dedicated to **African Liberation** through the study, comprehension, and **practical application** of **Science**, **Technology**, **Engineering**, and **Mathematics** (**S.T.E.M.**).

Eye Of Providence

Between the years of 2009 to 2014 I published seven (7) books covering topics on the Scientific Method, Mathematics, Electrical Engineering, Architecture, Thermodynamics, Computer Programming, and Aeronautics. It was my desire that the "**African Creation Energy**" series of books would motivate and inspire an **African Scientific Revolutionary Movement**. I have also observed that people are energized and galvanized into action through their form of **Spirituality** or **Religion.** Therefore, the working hypothesis (**speculation**/belief) of the "**African Creation Energy**" series of books has been to relate concepts, terms, symbols, and ideas from **Science**, **Technology**, **Engineering**, and **Mathematics** to concepts, terms, symbols, and ideas from traditional **African** culture, spirituality, and mythology in order to engineer an **African Spirituality** or **African Religion** centered around

Science, **Technology**, **Engineering**, and **Mathematics** which would in turn energize and galvanize African people and their descendents into the use and practical application of **Science**, **Technology**, **Engineering**, and **Mathematics** for the liberation of African people to be free of the dependence on others and develop agency and self-determination. This newly engineered African Spirituality would essentially be "*Science veiled in Mythology and Illustrated by Symbols.*" After covering a variety of topics over the years through the release of material in audio, video, and print media, I asked myself the question: "*What is the MOST important Science that African People should Learn and Master?*" After musing over this question for several years, I finally came to the conclusion that the answer is **SPECTROSCOPY**! This is not to say that other Sciences are not important, nor is it my intention to deemphasize the utility of Mathematics. On the contrary, extremely compelling, convincing, and persuasive arguments can be made for the importance and significance of every scientific discipline. And, Computer Programming is perhaps the most essential and fundamental skill needed by our generation in the modern economy. However, from my **perspective**, the study, comprehension, and practical application of the science of **Spectroscopy** will exponentially empower African Liberation and economic development by facilitating a means to identify and gain detailed knowledge about **Natural Resources**. The science of **Spectroscopy** will be explored in greater detail within this book, but, simply put; **Spectroscopy is the study of the interaction between Light and Matter**.

As the study of the interaction between light (electromagnetic radiation) and matter, **Spectroscopy** has an important symbolic significance to Science and Survival in general. The symbolic significance of **Spectroscopy** to Science and Survival is simply this: All of the information that you know, which you use to survive, has been obtained through the interaction between electromagnetic radiation and matter. Your 5 senses of Sight, Hearing, Taste, Smell, and Touch detect information about your environment and send that information to your brain in the form of electrical impulses. Everything you experience is the result of the interaction between electromagnetic radiation and matter! The information you gather by way of your experiences is then processed and utilized in the Scientific Method by the brain. Information and data is everywhere, and the faster you can gather and process information, the faster you can come to scientific conclusions to make informed decisions. If you had the ability to quickly scan your environment to determine the composition, benefits, and detriments of everything in the environment then you would be able to leverage the resources of Nature for your better good. Imagine being able to scan a surface before you touched it to ensure that nothing harmful is on the surface, or being able to scan food or drink before you consume it to determine the actual nutrient content of the meal, or being able to scan the air you breathe to ensure it is free of any toxins, or being able to scan a person to determine their DNA, or being able to scan the earth to find precious minerals – all of these abilities can be facilitated through the practical application of the Science of **Spectroscopy**.

The Symbolism & Significance of Spectroscopy to Science & Survival

The utility and practical application of **Spectroscopy** provides a unique solution to one of the essential challenges facing Africans and people of African descent worldwide. Although I am an African-American who was born and raised in the United States of America, I have been fortunate to travel to multiple countries across the African continent, and live and work with African people across the world. When I observe the condition of my people globally, I see that we make up the lower-class of any foreign society in which we find ourselves. And even within our own African societies, I see that we are disproportionately at an economic disadvantage as compared to societies of other cultures of people around the world. Some may argue that Africa is the richest continent in terms of Natural Resources; so why then after all of these years has this richness in Natural Resources not translated into wealth for the African people? Perhaps foreign exploitation, or internal corruption, or a combination of both is the explanation. Some may argue that African people are rich in Spirituality; so why then after all these years has this richness in Spirituality not translated into success for the African people? We are told that the wealth and success that we desire will come after we die and go to Heaven. Meanwhile, I observe other cultures of people experiencing Heaven right here on Earth. If Africa is the richest continent in terms of Natural Resources, then African people should be the richest people. If the various spiritual systems or religions that African people find themselves practicing, are not providing a means for African people to identify, harness, and leverage the richness of the Natural Resources of Africa to the benefit of African people, then an African Spirituality or Religion must be engineered which achieves this goal. It is my Hypothesis (belief) that a Spirituality or Religion centered around Science, Technology, Engineering, and Mathematics (S.T.E.M.), is the solution to the various problems, challenges, and issues that African people are faced with globally, and the science of **Spectroscopy** and its esoteric and exoteric applications is essential to this mission. The "**African Creation Energy**" series of books have demonstrated how African Spirituality can be synthesized with, and used to teach various topics in Science, Technology, Engineering, and Mathematics, and these books have been dedicated to integrating Science with African Spirituality for the purpose of the survival and general improved well-being of African people globally.

This book about the practical application of the Science of Spectroscopy for Survival and well being is entitled "**SPEC-RA-SCOPE**" because the term "**SPEC-RA-SCOPE**" is phonetically similar to the word "**Spectroscope**," which is a technological instrument used for producing and recording data in the Science of Spectroscopy. Also, the term "**SPEC-RA-SCOPE**" is formed by combining the three words "**Spec-**," meaning "*to observe*", **RA**, the name for the African Sun Deity from **Kemet (Ancient Egypt)** and the primary source of light here on Earth, and "**-Scope**" meaning "*to watch*". The formation of the term "**SPEC-RA-SCOPE**" reflects the idea of two **Eyes** (**Spec** and **Scope**) observing and watching the **light** from the **Sun** (**RA**), which serves as a good **mnemonic device** to remind **pupils** of the purpose and function of the **Science of Spectroscopy**.

The Symbolism & Significance of Spectroscopy to Science & Survival

Much like prerequisite courses a pupil encounters while matriculating through the education system, the main topic of this book builds upon concepts presented in my previous works. Therefore, this book will contain some information repeated from my other past books, lectures, and presentations which is relevant to the topic of this book and the science of **Spectroscopy**.

The title of each chapter of this book is a word containing the word "**spec**" meaning "**to watch or observe**," and thus this book can be considered a metaphorical "**Scroll of Eyes**". Symbols related to "eyes" and "observation" are used throughout this book because in Ancient Egypt in Africa, an "Eye" symbol was and is used as a composite symbol to represent all of the information received to the human mind by way of the senses. It is the information that we experience, with "Experience as the best Teacher" and us as the "**pupil**" (both student and **center of the eye**), which provides us with Science and a means to survive. This first chapter of this book is the **Prospectus**, which serves to "forecast" or describe the forthcoming book. Providence is defined as "*timely preparation for future eventualities,*" and so, the symbol of the "**Eye of Providence**" is used to represent the **Prospectus** chapter. The second chapter of this book gives my **Perspective** on the topics of Observation, Spectroscopy, and Science for Survival. In this book, I have intentionally written in the "first person singular" using the term "I" because it is phonetically similar to the word "eye." Additionally, in

Rastafarian culture, the phrases "I and I" is used to express oneness as a unity between "You and I" and "the divine and self" similar to the Hebrew phrase *"I am that I am"* or the Twi phrase from Ghana West African *"Me ne Me."* The format of every **African Creation Energy** book has been to show or establish a relationship between the scientific principles discussed to African cultural concepts and symbols, and this occurs in this book in the third chapter entitled "**Speculation**." The fourth and fifth chapters cover the sciences of **Spectroscopy** and **Spectrometry** respectively. The sixth chapter entitled "**Specifications**" provides the reader with hands-on projects and experiments to gain firsthand experience with the scientific principles discussed. The seventh chapter entitled "**Respect**" serves as an afterword providing context to the thoughts that served as the impetus of this project, and chapter eight entitled "**Inspector**" is about the author "**African Creation Energy**."

Spectroscopy is the study of the interaction between light and matter. The interaction between light and matter is what also occurs during observation. Observation is the first act in both **Science** and the **fight for survival**. So, in both the literal and figurative context, when it comes to the Science of **Spectroscopy**, "**It's on sight!**"

2. Per**spec**tive

The definition of the word "**Perspective**" means "***point of view***." The etymological origin of the word "**Perspective**" comes from the prefix "**per-**" meaning "***through***" and the root word "**-spec**" meaning "***to look or to observe***," which connotes the idea of "***looking through***". As the old adage says: "***the eyes are the window to the soul***," therefore, I would like to share my "soul" and perspective with you by inviting you to take a **look through my eyes to see my point of view**.

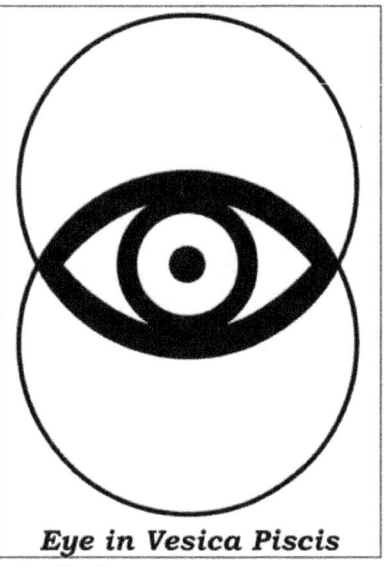

Eye in Vesica Piscis

The inability to view things and comprehend things from a different perspective is the source of many conflicts and debates. Of course, if someone feels wronged in a conflict, then they may desire to seek revenge if they live by the precept "***an eye for an eye***," and thus the conflict will perpetuate. If your perspective, and an opposing perspective, were modeled using two overlapping circles in a **Venn diagram**, then the common ground where there is shared comprehension of each perspective would be represented by the **Vesica Piscis** intersection of the two circles. Often times, we think that having different perspectives means that we can only arrive at **Relative Truths**, rather than the **Absolute Truth** (which is the **focus** of Science). This idea of individual perspective leading to "**Relative Truth**" is the sentiment expressed in the famous aphorism "***Beauty is in the Eye of the Beholder***."

However, a difference in perspective does not necessarily mean that each point of view is wrong. It may be necessary to literally observe the subject from a different position in order to comprehend the truth of a different perspective.

In Science, **Parallax Error** is a type of data recording observation error caused by a difference in **Perspective**. Parallax Error occurs when an object is viewed from different positions which distort the ability to obtain an accurate observation. Given the concept of Parallax Error, I wonder how many debates, arguments, fights, conflicts, wars, and instances of not "*seeing eye-to-eye*" could have been avoided if more time was spent attempting to "**see things from a different perspective**." In addition to information being received differently if obtained from a different position (both literally and metaphorically) it is possible that our sensory organs may be compromised with conditions like **blindness**, **partial blindness**, or **deafness**, which inhibit our ability to receive information fully or accurately. Even if we are not afflicted with one of the aforementioned conditions, there are conditions like **scotoma** which is a partial loss, blind spot, or alteration to normal **vision** which both literally and metaphorically affects the way we "**see things**." And it is important to distinguish the difference between **sensation** and **perception**. **Sensation** is the information we receive from our

Example of Parallax Error

five senses of sight, sound, smell, touch, and taste, to our brain. **Perception** is the way our brain interprets or "**makes sense**" of the information from our five senses. So even if your sensory organs send accurate information to your brain, it is possible for your brain to misinterpret the information and "**perceive**" the information incorrectly. For example, an **optical illusion** occurs because of the brain's misinterpretation of **visual** stimuli causing **perception** to differ from **reality**. There is an old aphorism that says "*believe half of what you see and none of what you hear*," however in this day and age of **doctored photos, holograms, augmented reality**, and "**deep fakes**", I think a better precept to adopt is "*believe nothing and question everything* (including this statement)."

"**Consciousnes**s" is the bridge between **Sensation** and **Perception.** "**Consciousness**" is the state of being aware, and occurs after information is received, and prior to - and in conjunction with - information being interpreted. As the connection between sensation and perception, consciousness is usually represented by the "**third eye**" symbolic of the mind. Most people think that **consciousness** and **wakefulness** are synonyms. That is to say, we think that to be **awake**, is also to be **conscious**. While it is most common for consciousness and wakefulness to occur simultaneously, in **cognitive science** however, these two terms have important distinctive definitions. Consciousness is the state of being **awake and aware**, able to **perceive, receive,** and **process stimuli and information from one's environment**. When you go to sleep, this is an **altered state of consciousness**, with limited, to no ability, to perceive, receive, and process stimuli and information from one's environment. When Neuroscientists study the EEG brain waves of a sleeping person, they find that during a night's sleep, a portion of the time is spent in the waking state, even though the person is not fully conscious. **Parasomnia disorders** such as "**sleep walking**" or "**sleep talking**" are examples of instances where a person is in a "waking" state, but not fully conscious. **Daydreaming** is another example of a mental state where a person is awake, but not conscious of their immediate surroundings. Conversely, **Sleep paralysis** is a condition where the mind is awake and conscious, but the body is not awake and unable to become active. What we can take away from all this is that: 1) Consciousness and wakefulness commonly occur seemingly simultaneously; 2) Consciousness also requires one to be awake; 3) It is possible to be awake but not conscious; and 4) It is possible to be conscious and mentally awake but not physically awake and active. These findings from cognitive science about **consciousness** and **wakefulness** are also applicable to the use of the terms "**Conscious**" and "**Woke**" in the Black community. The term "**Consciousness**" in the **Black community** has a long and storied history throughout Africa and the African Diaspora, stemming back to the early 1900s, and has to do with an awareness of one's black identity, and nonconformity to mainstream social, political, economic, religious, and spiritual constructs. The **UNIA** (Universal Negro Improvement Association), **Moorish Science Temple**, **Nation of Islam**, **5 Percenters**, **Hebrew Israelites**, **Nuwaupians**, **Black Panther Party**, **SCLC** (Southern Christian Leadership Conference), and **BLM** (Black Lives Matter), are all examples of **black conscious movements** in America. In recent years, the term "**Woke**" has surfaced as an idiom essentially referring to the same concepts, precepts, and principles as "**Conscious**", but with more of a focus on social, political, and economic awareness. In recent years, the term "**Conscious**" has become associated more with a focus on historical, cultural, religious and spiritual awareness. The Activism of someone who labels them self as "**Woke**," tends to be of a social, political, and economic nature, whereas, the activism of someone who labels them self "**Conscious**," tends to be of a historical, cultural, religious or spiritual nature. If we were to retrospectively apply the new definitions and connotations that the terms "**Woke**" and "**Conscious**" have taken on in recent years, to the aforementioned groups, then we could classify the **UNIA**, **Black Panther Party**, **SCLC**, and **BLM** as "**Woke**," and the **Nation of Islam**, **5 Percenters**, **Hebrew Israelites**, and **Nuwaupians** as "**Conscious**". In some regard, "Woke," seems like a re-branding of "Conscious". But if "**Woke**" has become used to refer to more social, political, and economic awareness, and "**Conscious**" has become used to refer to more historical, cultural, religious and spiritual awareness, then just like in cognitive science, it is most common to be simultaneously "**woke**" and "**conscious**", that is to say, having simultaneous affiliation and interest in organizations concerned with both social, political, and economic issues as well as historical, cultural, religious and spiritual issues. Also, just like the concept of **Sleep walking** and **Sleep Talking** in cognitive science, it is possible to be **"Woke" but not "Conscious"**, that is to say, have affiliation with, and interest in, organizations primarily concerned with social, political, and economic issues, and having no affiliation with, or interest in, organizations concerned with historical, cultural, religious and spiritual issues. These individuals are aware of the social injustices in the world, but have no knowledge of their historical past or traditional systems of spirituality. And lastly, just like the concept of **Sleep Paralysis** in Cognitive Science, it is possible to be **"Conscious" but not totally "woke"**, that is to say, having interest in historical, cultural, religious, and spiritual issues, and having no interest in social, political, and economic issues. These individuals are fully aware of their historical past, have "**knowledge of self**", and practice some form of traditional spirituality, but have no concern or activism in regards to the social, political, or economic injustices in the world. Essentially, "**Woke**" and "**Conscious**" are mental states, and the ultimate goal is to become **Active**, with an expression of one's awareness demonstrated through **practical application**. If the mind is represented by the "**third eye**," then "Conscious" and "Woke" should be represented as the two eyes (two hemispheres of the brain) which come together to form **Binocular Vision** and perceive reality.

There is a famous colloquial proverb that states "**Perception is Reality**." Comprehending the difference between **sensation** and **perception**, then we can say that if your ability to receive sensory information from your environment is not compromised, then it may be possible for you to **sense** reality, but your ability to **perceive** reality is dependent on the mind's ability to **reason** or "**makes sense**" of **sensory information**. If your sensation is skewed, or your perception is skewed, then what you perceive as reality will also be skewed. So due to the fact that each person may have different perceptions of reality, then it is safe to assert that "**alternate realities**" do exist, at least within the mind of each person. It is important to disambiguate between the terms **Reality**, **Universe**, and **Dimension**, because these ideas and terms are sometimes conflated and used incorrectly interchangeably. In physics, the term "**Universe**" refers to all of space, time, energy, and matter, including all planets, stars, galaxies, people, animals, and subatomic particles. The term "**Reality**" is a philosophical concept referring to the **perception of existence**, or the **perception of the universe**. Due to the fact that perception can vary, then it follows that one's perception of reality can also vary. It is possible to alter your perception of reality, or enter into an **alternate reality**, through the use of psychoactive substances such as drugs and alcohol. Additionally, through the use of technology, it is also possible to enter into alternate realities through the use of **Virtual Reality** or **Augmented Reality** technologies. In physics and mathematics, the term **"Dimension"** is used to define the number of coordinates needed to define an object in space. Through the mathematics of **Geometry**, we are very well acquainted with the concept of dimensions. A line is a 1-dimensional object, a square is a 2-Dimensional object, and a cube is a 3-dimensional object. In classical physics, space is described in 3-dimensions, or 3D, length, width, and height, which are the degrees of freedom, or basic directions in which we can move: up/down, left/right, and backward/forward. If you have ever done geometric mathematic equations to solve the area of a square, or the area of a cube, or the volume of a pyramid, then you have literally performed **multi-dimensional mathematics**. Mathematicians also frequently solve equations involving higher dimensional objects. Some examples of higher 4-dimensional objects include the tesseract, pentatope, hexadecachoron, octaplex, dodecaplex, and tetraplex. To summarize, **Dimensions** define **space within a Universe**, and **Reality** defines **perception of mind within a universe**. It is possible to have multiple dimensions within one universe, and it is possible to have multiple realities within one universe.

In **Physics**, there is a famous experiment called the "**Double-slit experiment**" which was performed in the early 1800s to demonstrate the **dual nature of light** as both a particle and a wave (called **wave-particle duality**). The fascinating outcome of the "**Double-slit experiment**" is that when observing the experiment, light behaves as a wave, but when no one is observing the experiment, light behaves as a particle. One of the conclusions from the "**Double-slit experiment**" is that "**reality does not exist until it is observed**." As the American theoretical physicist **John Wheeler** once said, "*Reality is made of information, and information is created by observation.*"

The idea that observation defines reality is expressed in the "**Schrödinger's cat**" thought experiment in that given a cat in a box, you will not know the reality of the state of the cat as alive or dead until you open the box and observe the cat. The "**Schrödinger's cat**" thought experiment provides a good analogy for understanding the idea of **Quantum Superposition** of reality existing in multiple states until it is observed. Not only does observation define reality, it has been scientifically demonstrated that the **amount of observation** can affect reality. In February 1998, scientists at the **Weizmann Institute of Science** conducted a version of the "**Double-slit experiment**" using an electronic detector as the "**observer**" of the experiment and noted that as observation increased, the observer's influence on the experiment also increased citing that "*the mere act of observation affects the experimental findings (reality)*." One simple way of understanding and thinking about this concept in laymen's terms is: "*You act different when people are looking at you.*" Some people have misinterpreted this finding as meaning that "**consciousness defines reality**," however in the case of the electronic detector used as the "**observer**" in the experiment, the device itself was not "**conscious**." A more appropriate interpretation is that "**sensation in the form of observation defines reality**". The idea that observation affects reality is expressed in Physics as the **Observer Effect**. Comprehending the way we see any object is by light having to reflect off of the object then enter our eyes, we have to comprehend that when the light touches the object, the object is changed (even if by a little bit) and thus the act of Observation affects Reality. The **Observer Effect** from science is speculated as a possible explanation to the psychic phenomena of **Scopaesthesia**, also called **Gaze detection** or **Gaze perception**, which is the thought or idea that you can "**feel people looking at you**" or "**feel it when someone is watching you**" or when it feels like "**all eyes are on me.**"

So when we put into **perspective** the fact that we observe nature in order to comprehend reality, but the act of observing affects reality, and our ability to accurately observe or accurately comprehend what we observe affects our perception of reality, then this underscores the importance of applying a rigorous **Scientific Method** as a **way of life**. It is the practical application of the Scientific Method which provides us with **Discernment**, the ability to judge between right and wrong. Much like **Metacognition** is the act of being aware of one's own thoughts, Scientists must be aware and willing to constantly re-evaluate any underlying assumptions that may lead to inaccurate conclusions. In Psychology, the **Mere Exposure Effect** occurs when people develop a preference for things that they are familiar with and have an aversion toward things that are unfamiliar. Therefore, it is possible to be bias towards the things you already know or believe and reject new information that is unfamiliar which you do not know. The bias of the **Mere Exposure Effect** can be compounded with the psychological phenomena known as the **Illusory Truth Effect** where people believe information is true after repeated exposure. For example, if someone is repeatedly told a lie that they know is a lie, over time due to repeated exposure, the **Illusory Truth Effect** states that the person may begin to believe the lie as truth. The **Illusory Truth Effect** occurs in **Gaslighting** which is a form of **psychological manipulation** where you are made to question the truth of your own memory and perception of reality. With multiple competing and contradictory versions of reality occurring in the mind, a state of **Cognitive Dissonance** forms in the mind which breeds ideological paradoxes and hypocritical actions until the individual takes actions towards reconciliation. The **Mere Exposure Effect** and the **Illusory Truth Effect** illustrate how **apperception** can give birth to misinformation and **pseudoscience**. **Apperception** is when you perceive new information based on old past information. However, if the old past information is wrong or incorrect, then **apperception** can lead to the new information being perceived incorrectly. Since Science is the systematic study of nature, then everything in nature can be studied scientifically, including that which we qualify as **pseudoscience**. By **scientifically studying pseudoscience**, there are some interesting outcomes discovered through the investigation. It is incorrect to think that pseudoscience is not based on evidence and not based on data. If pseudoscience was not based on any evidence or data, then it would be impossible to argue with someone who was promoting pseudoscientific ideas because they would have no information (data) to discuss. If you are ever in a debate with someone who is arguing a pseudoscientific point, and you are presenting evidence to support your scientific claims, the person arguing the pseudoscientific point will also be able to retort with evidence to support their pseudoscientific claim. Promoters of pseudoscientific claims are capable of citing and presenting vast amounts of data and information to support their argument. When a person is promoting a pseudoscientific claim, they may not necessarily be relying on **pseudo-information** (false information) in order to support the pseudoscientific claim. It is possible for a pseudoscientific claim to be supported by accurate and correct information. It is important to distinguish the difference between **pseudoscience** and **pseudo-information**. The word "**pseudo-information**" is just another way of saying misinformation, false information, or incorrect information.

In English, the six words of scientific investigation which precede any question are **Who, What, Where, When, Why**, and **How**. The answers to these questions will always give an answer that is one of the four aspects of Nature, namely:

Space, **Matter**, **Time**, and **Energy**. The questions "Who and What" inquire into the Matter aspect of Nature, the question "Where" inquires into the Space aspect of Nature, and the question "When" inquires into the Time aspect of Nature. The answers to the "Who, What, Where, and When" questions are descriptive and provide the individual with a basic semantic definition and identification about information. The questions "Why and How" inquire into the "**Energy**" aspect of Nature as they seek to find the **cause**, method, or **explanation** by which some experienced information occurred.

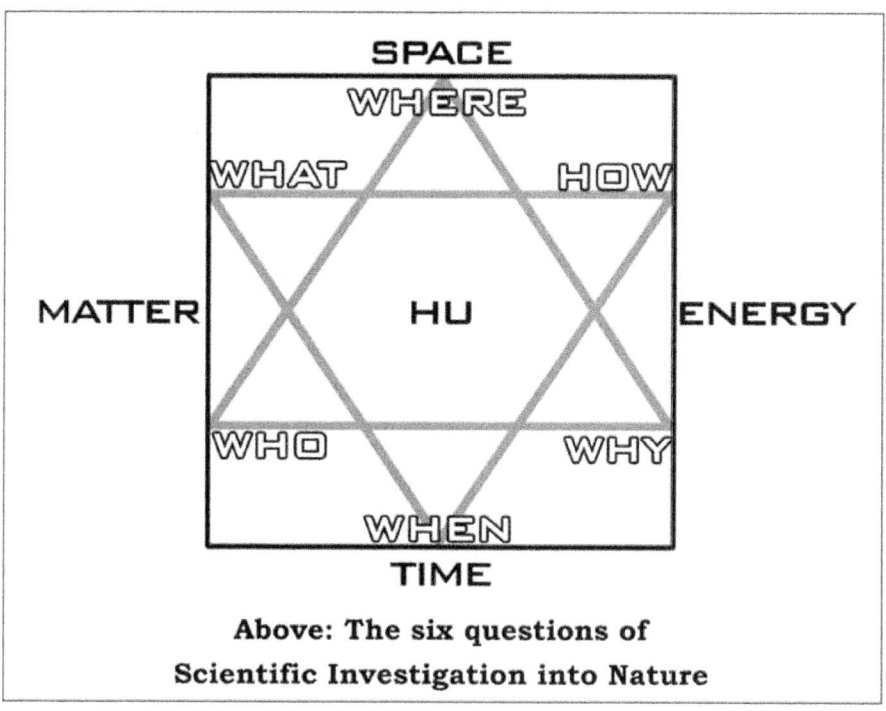

Above: The six questions of Scientific Investigation into Nature

The answers to the "Who, What, Where, and When" questions, where the answer is descriptive about either "Space, Matter, or Time," is what produces **Scientific Laws**. The answers to the "Why and How" questions, where the answer is explanatory about "Energy," is what produces **Scientific Theories**. A **Scientific Law** is a description of an **Observed** phenomena in

Nature often expressed in a single mathematic equation. Scientific Laws are cornerstones of Science and do not change. In E.A. Wallis Budge's **"Egyptian Hieroglyphic Dictionary Vol. 1"** he cites that in the Ancient Egyptian language of Medu Neter, **Scientific Laws** were called by the term "*Hepu n Rekh.*" A **Scientific Theory** is an in-depth explanation of the "how" and "why" for an observed phenomena in Nature, and may make reference to one or more **Scientific Laws**. For example, **Ohm's Law** is one of many **Scientific Laws** which make up the **Scientific Theory** called "**Electrical Circuit Theory**." Scientific Theories may change or be updated over time if better explanations are discovered. **Pseudo-information** is incorrect answers to the "Who, What, Where, and When" questions. **Pseudo-information** can be thought of as a perversion of **Scientific Law**. **Pseudoscience** is an incorrect methodology in answering the "Why and How" questions. **Pseudoscience** can be thought of as a perversion of **Scientific Theory**. **Pseudo-information** is incorrect information and **Pseudoscience** is incorrect explanation.

Pseudo-information (false information) is easily refuted by correct information. One of the reasons why people accept **pseudo-information** (false information) is because they have learned overtime to accept and trust information coming from a particular source. If the source of information is a particular person, people may have learned to trust and rely on the person due to the person's ability to demonstrate some proficiency in a particular area. However, if the trusted source of information ever gets the information wrong, people may accept the wrong information (pseudo-information) because they have learned over time to **trust** and **believe** in the source of information. In the book entitled *"The Misinformation Age: How False Beliefs Spread"* by philosophers of Science Cailin O'Connor and James Owen Weatherall, they make the point that *"the acceptance and promotion of misinformation is more about the source you trust, than it is about what you personally think."* It is accepting, trusting, and believing in **pseudo-information** (false information) which leads to people being scammed and conned. The word "**con-man**" is short for "**Confidence Man**," and refers to the fact that in order for someone to scam or "con" someone else, they must first gain the person's confidence that they can be trusted and believed. The confidence and trust is gained by presenting correct and accurate information, and the scam comes when misinformation or **pseudo-information** (false information) is presented. The source of information that we have learned to trust may not be another person, but it may be our own senses (sight, hearing, touch, smell, and taste) which can also be prone to error and occasionally collect **pseudo-information** (false information). **Pseudo-information** (false information) is problematic, but is easily refuted and corrected with facts and accurate information. However, it is possible for a **pseudoscientific** claim to be based on correct information (evidence). It is also possible for a **pseudoscientific** claim to be formed by a logical and reasonable argument. There are different types of Reason including **Inductive Reasoning**, **Abductive Reasoning**, and **Deductive Reasoning** just to name a few. **Inductive Reasoning** is the mental process of making **generalizations** from specific experiences. It is possible for conclusions arrived at through Inductive Reasoning to be true or false. If the premises of an Inductive Reasoning argument are true, then it is probable that the conclusion is true. **Abductive Reasoning** is the mental process of making **explanations** for experiences. The conclusions of Abductive Reasoning arguments must be plausible, but are not uniquely the only explanation and thus contain some degree of uncertainty. **Deductive Reasoning** is the mental process of determining **truth** by deriving specific logical conclusions from ideas and concepts in the mind. The conclusions of Deductive Reasoning arguments are guaranteed to be true if the premises are true. Deductive Reason is "**Sound Right Reason**" in that if a Deductive argument uses valid logic, and correct premises ("**Right Knowledge**"), then the argument is referred to as "sound," otherwise it is "unsound."

Inductive Reasoning, Abductive Reasoning, and Deductive Reasoning occur in a cycle in scientific thinking in that:

1. Inductive Reasoning is used to conceptualize generalizations (rules)

2. Abductive Reasoning is used to develop explanatory causes

3. Deductive Reasoning is used to make specific predictions

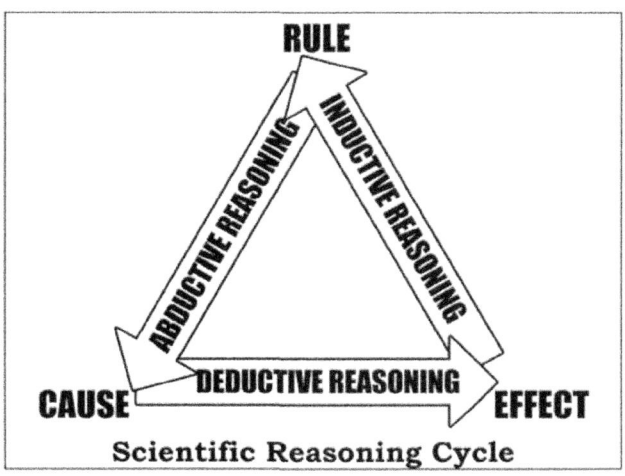

Once the different forms of Reason are comprehended and understood, it can be observed that **pseudoscience** does utilize the Inductive and Abductive forms of reason. What makes **pseudoscience** so deceptive is that just like science;

25

pseudoscience is based on **Evidence, Experience**, and **Reason**. **Pseudoscience** is defined as *"a collection of beliefs or practices mistakenly regarded as being based on the **scientific method**."* If the **pseudoscientific method** can be mistaken for the **scientific method**, then we must examine both methodologies in order to discern science from pseudoscience.

In the book "**The Science of Sciences and The Science in Sciences**" by **African Creation Energy**, the **Scientific Method** and the **Mathematic Method**, or the **Empirical Method** and the **Rational Method**, are discussed in terms of a cyclic process with the eight stages being **Experience, Comprehension, Theories, Analysis, Hypothesis, Synthesis, Experiment**, and **Results**. Within these eight stages, the point where the **Scientific Method** transforms into the **pseudoscientific method** is at the point of Experimentation. The stage of Experimentation is the stage dedicated to data collection and information gathering. During the stage of Experimentation, a **Scientist** will gather ALL of the information whether the information supports or refutes the Hypothesis. A **Pseudo-scientist** will also actually go out and gather evidence, information, and data during the stage of Experimentation, but will only gather evidence, information, and data in support of their hypothesis, and will ignore evidence, information, and data which refutes or disproves their hypothesis. In the Abrahamic religious traditions, it is said that there are **7 Archangels plus one**, the **Devil** or **Satan**, who used to be one of the highest Angels before he rebelled and was kicked out of heaven. If the eight stages of **Experience, Comprehension, Theories, Analysis, Hypothesis, Synthesis, Experiment**, and **Results** were to be personified, and an analogy were to be made between these eight stages to the Archangels from the Abrahamic religious traditions, then when the stage of

Experimentation transforms the **Scientific Method** into the **pseudoscientific method**, this can be likened to the fallen Angel who was **Satan**. The word and name "**Satan**" means "*adversary, opposition, or one who seeks to seduce humans into falsehood*," and just like "Satan," pseudoscience seduces humans into falsehood. The **pseudoscientific method** makes us stray off of the **true path** of Science; metaphorically called the **Sirat al-Mustaqim** in the religion of **Islam.**

Scientists allow the data and evidence to guide their conclusions. However, rather than allowing the data and evidence guide them, Pseudo-scientists allow their own hypothesis and beliefs to be their guide, and filter information so that only evidence is collected which supports their beliefs. Given the many ideas, assumptions, beliefs, and hypotheses that the mind generates to comprehend and explain nature, when observations are made to test the hypotheses, often times you will obtain data and information in support of the idea and you will find data and information that disproves the idea. In the cases where there is evidence on both sides, supporting and refuting the idea, then **Statistical Reasoning** must be employed to determine which conclusion is most probable and most likely. This can be metaphorically symbolized as "weighing the information on a scale" to see which side has more evidence to support it. On the way to determining the truth and validity of an idea, you must past through the three steps of **Possible, Plausible,** and **Probable**. Possible refers to information that has the potential to be true, but also may not be true. There are almost an infinite number of possibilities that can be imagined for any given subject. As human beings, we have neither the time nor the energy to consider every single thing that as the mere possibility of being true. So, we use Plausibility to determine which Possibilities are most reasonable. Plausible refers to information that is determined true by logic and reason. From the infinite number of possibilities, we use the different forms of logic and reason to determine what is most plausible. Everything determined as Plausible is also Possible, but not everything Possible is Plausible. Plausibility can be determined by way of reason "*a priori*," before specific evidence and data is collected. This relates to the old aphorism, "the absence of evidence is not evidence of absence." While it is true that the absence of evidence is not necessarily evidence of absence; that is to say, just because there is no empirical proof of something does not mean that empirical proof does not exist. However, until information that is proven by reason has empirical evidence, then Plausibility is the best qualification that can be given in the absence of evidence. Ideally, you want to have evidence, facts, and data to support your plausible argument. When unbiasedly collecting evidence and data, in most cases you will find evidence both supporting and refuting your idea, and using statistical probability facilitates the ability to determine what is most likely or most probable. Probable refers to information that is most statistically likely to be true. Everything Probable is Plausible, but not everything Plausible is Probable. Probability is determined by way of statistical reasoning when weighing the supporting evidence versus the refuting evidence. When the evidence supporting an idea is greater than the evidence refuting an idea in a ratio greater than 50%, then it is fair to say that the argument is "probable" (i.e. most likely to be true). However, within probability there are various degrees of strength, from greater than 50% to 99.9%. Something that is 75% likely to be true is better than something that is only 51% likely to be true, and something that is 85% likely to be true is better than something that is only 75% likely to be true, and so on. When you get to the level of 90%, 95%, and 99% probable, then you are in the range of "Statistical Significance," where there may be some refuting evidence, but the overwhelming majority of the evidence shows that the idea is highly likely to be true. When people argue against ideas that are true by way of "Probability," they attempt to find the small amount of information in the minority that refutes the idea, and ignore the overwhelming majority of information that supports the idea. Choosing to only accept certain information and willingly ignoring other information is the methodology of Pseudoscience and not Science. For most ideas and concepts, there will be information supporting and refuting the idea, and being able to say something is most likely, most probable, or statistically significant is the best qualification that can be given to the majority of concepts. Of all the concepts and ideas out there, the concepts and ideas that are determined to be 100% probable and 100% certain with no refuting evidence are actually in the minority.

Pseudo-scientist only seek out information which proves themselves right, and ignore information which proves them wrong. Pseudoscience is so deceptive because evidence can actually be presented to suggest that the pseudoscientific claim is correct, but the Pseudo-scientist has failed to research or present evidence about why the pseudoscientific claim is incorrect. The **Scientific Method** becomes the **pseudoscientific method** when **Confirmation Bias** corrupts the collection of evidence and data. **Confirmation Bias** occurs when an individual actively seeks out evidence, information, and data which confirms and reinforces their preexisting beliefs. The **pseudoscientific method** is rooted in a fear of **falsification**; the fear of being wrong.

Science relies on **falsification** in order to determine if a hypothesis is correct. **Falsification** is the act of proving something wrong. When evidence is gathered to **falsify** or disprove an idea, and no evidence, or very little evidence, can be obtained to disprove the idea, then it is fair to say the idea is correct or statistically likely. When all the sources of evidence, information, and data have been exhausted to disprove an idea, and the idea cannot be disproven, then you can have confidence in the accuracy of the idea. In the **pseudoscientific method** where only evidence is gathered in support of the idea, then you do not know if evidence exists which disproves the idea, so it is not safe to conclude that the idea is correct. When you practice science, it is a very **humbling endeavor**, because you are actually wrong more often than you are right. Your mind tends to generate more ideas, assumptions, and hypothesis about the way the world works than are actually correct; so you are often proven wrong. But, when you are right using the **Scientific Method** you can have confidence in being right. It is possible to be right using the **pseudoscientific method**; however you cannot have confidence in the accuracy of the idea because no effort was made to disprove the idea. So if you are right using the **pseudoscientific method**, it means you just got lucky. **Pseudoscience** creates a "**win-win scenario**" in that by using the **pseudoscientific method** you will "win" by finding evidence to support your idea, and you will not "loose" by looking for evidence which disproves or falsifies your idea. Because of the "**win-win scenario**" created by the **pseudoscientific method**, it appeals to the individual's **ego** to "**want to be right**" and not have your feelings hurt by being wrong. At some point in time in everyone's life we all fall a victim to the **seductiveness** of the **pseudoscientific method** which "**looks for evidence to prove that you are right**."

Even within the professional and corporate world, the **pseudoscientific method** shows up in the curation of data; that is **picking and choosing** what data to present and how to present it in a way that is **favorable** to a company's bottom line. The **pseudoscientific method** is pervasive and shows up in debates, conspiracy theories, racism, and the legal system, just to name a few instances where it manifests in life. The **pseudoscientific method** of looking for and presenting information which supports your belief, opinion, or claim is the method that is taught in language arts courses in schools around the world for writing and constructing a **Persuasive Argument**, which is why so many people do not intuitively see why the **pseudoscientific method** should not be used to determine the accuracy of an idea. The danger in using the **pseudoscientific method** to determine the accuracy of an idea is that there are many things in Nature which operate counter intuitively to what we may think, and so using the **pseudoscientific method** where we attempt to prove ourselves right will ultimately lead to not gaining a true understanding of Nature. It is possible to follow the **Scientific Method** and yield a wrong result and that is perfectly fine because it adds value to the body of knowledge that whatever hypothesis failed is not an accurate explanation for the observed phenomena and should be eliminated as an explanatory option. What you learn by following the **Scientific Method** is that whenever you are wrong, whenever you are falsified, it gives you the opportunity to learn through your failure. In athletics, it is said that "**you have to learn to lose before you learn to win**" and the same is true when applying the **Scientific Method**. The idea of being "**all wise, right, and exact**" or **Omniscient** is a logical fallacy. The **Omniscience logical fallacy** is that if a being knows everything, this means the being does not know nothing, and if the being does not know nothing, then there is something that is unknown, therefore the being cannot know everything. Even the **God** in the **Bible** made mistakes, as it states in the book of **Genesis 6:6 "<u>The Lord regretted</u>."** Or when the **God** in the **Bible created light** in the book of **Genesis 1:3** and "**God saw that the light was good**" which means there was a possibility that the light may not have been good (i.e. an undesired outcome). So the point is that even though the **pseudoscientific method** gives you the impression that you are right and feeds your ego, it robs you of the opportunity to learn from your mistakes and have a true and accurate understanding of Nature. Because the **pseudoscientific method** feeds the ego, and the **Scientific Method** forces you to be humble, two distinct personalities manifest in individuals who are practitioners of each method. Practitioners of the **pseudoscientific method** tend to be egotistical, proud, boastful, charismatic, confident, loud, obstinate, and speak in terms of "absolutes". Practitioners of the **Scientific Method** tend to be humble, inquisitive, amenable, and speak in terms of probabilities and likelihoods. Human beings are naturally attracted to leaders with characteristics that the **pseudoscientific method** breeds. The appeal of the **pseudoscientific method** itself, as well as the natural inclination to follow leaders with characteristics bred by the **pseudoscientific method** can be explained by **Evolution** through **Science**.

As previously mentioned, when you study Nature, you find that there are many counterintuitive things that exist within Nature. One thing that is counterintuitive is the instances where the **pseudoscientific method** is actually practical. Intuitively, if the **Scientific Method** yields results that are accurate, then you would think that employing the **Scientific**

Method would be the best approach all the time. However, there are instances where using the **Scientific Method** can be detrimental, and using the **pseudoscientific method** can be beneficial. Recall that the **Scientific Method** seeks to find evidence which falsifies the idea to determine if the idea is correct, whereas the **pseudoscientific method** seeks to find evidence which confirms the idea to determine if the idea is correct. Now, imagine two ancient ancestors, walking in the jungle at night, and they hear a sound behind a bush that sounds like a carnivorous animal which may try to kill and eat them. If one ancestor were to use the **Scientific Method** and try to falsify the hypothesis that the animal is indeed dangerous, and go investigate behind the bush, then the ancestor may end up being killed by the animal. However, if the other ancestor were to use the **pseudoscientific method** and look for any evidence which confirms the assumption that the animal is dangerous without trying to falsify that assumption by investigating behind the bush, then the ancestor would have survived another day by running away when the sound was heard. Some people refer to this colloquially as the "*paralysis in the analysis*," that is to say, spending so much time analyzing and being scientific that you become unable to take action. So in this instance, using the **Scientific Method** to falsify hypotheses was detrimental, and using the **pseudoscientific method** led to **self-preservation** and the genes of the ancestor who lived by the **pseudoscientific method** got passed down to us through the generations through evolution. In their paper entitled "*Conspiracy theories: Evolved functions and psychological mechanisms*," the authors Van Prooijen and Van Vugt call this phenomena "**Adaptive Conspiracism Hypothesis**", where human beings evolved the tendency to look for and accept any evidence which may indicate a hostile condition in order to avoid the danger and stay safe.

In the story of Adam and Eve in the book of Genesis in the Bible, God told Adam and Eve not to "eat the fruit from the tree of knowledge, for in the day you do so, you shall surely die." Then the serpent came to Eve and told her that if she ate from the tree she would not die. Now, Eve has two competing hypotheses: 1) if she eats from the tree, then she will die, and 2) if she eats from the tree she will not die. With these two competing hypotheses, Eve performs the experiment and takes a bite of the fruit. Eve uses the **Scientific Method** and attempts to falsify the hypothesis that the fruit is dangerous. After using the **Scientific Method** and biting the fruit, Eve finds out the fruit is not dangerous and that the serpent was correct and God was not correct. Of course in the Biblical story, this instance of Eve using the **Scientific Method** leads to God kicking Adam and Eve out of the Garden. But in this story of Adam and Eve in the Bible, the character of Eve jeopardizes her well-being and potentially compromises her own **self-preservation** by performing the experiment to take a bite of a fruit that could potentially kill her.

One famous idiom states the "**Self-Preservation is the first Law of Nature**." **Self-Preservation** is the process where an organism prevents itself from being harmed or killed. The Human being will use either the **Scientific Method** or the **pseudoscientific method** if the result is **Self-Preservation**. In cases where using the **Scientific Method** to falsify hypotheses can lead to the individual being harmed or killed, then the individual will instinctively use the **pseudoscientific method** in order to preserve one's self. Just like in some instances **mould** can be harmful, and in other instances mould can be refined to make **penicillin** medication which can be beneficial, there are some instances where utilizing a **pseudoscientific** thought process is the best approach for self-preservation. Your sense of "self" not only refers to your physical self, but also refers to your "**mental**" self, and the various ideas, concepts, and beliefs that have come together to form your mentality, personality, and identity. **Self-Preservation** also refers to the **Defense Mechanisms** utilized to protect one's mental schema. If new information is experienced which disproves one of the ideas, concepts, or beliefs that form an individual's "mental" self, then the individual may instinctively utilize the **pseudoscientific method** for mental **Self-Preservation**. As the old proverbs state, "**ignorance is bliss**" and also "**the truth hurts**." This is why when debating certain topics with certain people, you may hear someone say "*you are attacking me*," because the topic of debate has become a part of the individual's "mental" self schema, and they cannot separate them self from the topic. Disproving the topic would destroy part of who the individual is in their mental schema. **Pseudoscience** is a lie we tell ourselves, and it is possible for entire Nations to be built around these lies. Terms like a "**Noble lie**," a "**Pious Fiction**," and most recently "**Alternative Facts**" and "**Truth Adjacent**," all refer to misinformation (Pseudo-information) or Pseudoscience that a Nation uses to galvanize its people. While the **Scientific Method** is used to build the self and build and construct the Nation, the **pseudoscientific method** is used to maintain the sense of self and maintain the "*status quo*." For every historical empire, kingdom, or great culture that you can think of, there was some form of **Mythology** utilized as a tool to control the masses of people. When we study the science of Anthropology, we find that historically

Human beings have been naturally fascinated and attracted to Mythology. The challenge with creating and establishing a Nation based on Science is reconciling the Human desire for mythology with the purpose and objective of Science to provide explanations grounded in reality and truth. From my perspective, there are two approaches of reconciling this dilemma. One way of reconciling this dilemma is by having a Nation founded and based on Science and providing the people with mythology in the form of Science Fiction as purely entertainment. This can be observed in the modern day where people are galvanized around their favorite Science Fiction movies, they dress up like the characters, and collect the figurines, but they know that the object of their fascination is not real. This allows the people to have their Science and have their Mythology, and not be confused or assume that the Mythology is real. Another method to reconciling the dilemma of having a Nation founded and based on Science yet still satisfying people's desire for Mythology is the approach which seems to have been employed by the Ancient Egyptians and various African and Indigenous cultures across the world through time, and that is to craft a Mythology that has interpretations and explanations related to scientific concepts. In the book "Stolen Legacy" by George G.M. James, it states, "*It was the method of the Egyptian to conceal the truth by use of myths, parables, magical principles, number philosophy, and hieroglyphics as a primitive scientific method.*" Additionally, the **Freemason** secret society has followed in the tradition of the Ancient Egyptians with their "*system of morality, veiled in allegory, and illustrated in symbols.*" According to the field of **Ethnoscience** which studies how various cultures of people develop different forms of knowledge and beliefs, it has been the practice of the Western world (Europe and America) to separate science and spirituality, whereas traditional indigenous cultures tend to have a holistic worldview where science and spirituality are interconnected and thus mythology can have both scientific and spiritual implications. The word "**Math**" and the word "**Myth**" both come from words meaning "**thought**", so if **Mathematics** can be used to encode scientific concepts into abstractions, then **Mythology** can be used to serve the same purpose. Both Math and Myth involve the process of abstracting the underlying essence of a concept and generalizing it into a symbolic abstraction. **Ethnomathematics** is the study of the different forms of mathematics (**symbolic thinking**) used among different cultural groups. In a Nation founded on Science, Mythologies with Scientific interpretations will serve the dual purpose satisfying people's desire for mythology and simultaneously providing the people with scientific information. While the general sense of the word "myth" gives rise to ideas that are not entirely true, a mythology or allegory engineered to have scientific interpretations and provide scientific explanations should be viewed to a symbolic abstraction akin to Mathematics and not dismissed as **Pseudoscience**.

However, where the **pseudoscientific method** may work in some instances for self-preservation, if the "self" is ever compromised and gets sick, or needs to be healed, it is the **Scientific Method** which leads to **self-improvement**. The **Scientific Method** is used to create, to develop, and build. You utilize the **Scientific Method** for your food, clothing, shelter, and medication. This is why we see the earliest example of the **Scientific Method** on a medical papyrus from **Ancient Egypt**. The **Scientific Method** forces the self to grow, update, change, and progress. It is through the scientific investigation of the concept of Pseudoscience that we are able to identify instances where there is value in using and applying Pseudoscience. The **Scientific Method** and the **Pseudoscientific Method** essentially act as forms of **technology** to be used and applied when needed. **Technology** refers to tools, machines, techniques, skills, **methods**, and processes used to accomplish some objective. **Technology** is produced by **Engineering**, and Engineering is applied **Science**.

The **Scientific Method** is the technology that underpins all science and is used to accomplish the objective of Science. The objective of Science is the systematic study of the natural world gained through observation and experimentation. **Science**, **Engineering**, and **Technology** work together in a cyclic fashion. New scientific findings lead to new applications in Engineering which leads to new technology which lead to new scientific findings, and the cycle continues.

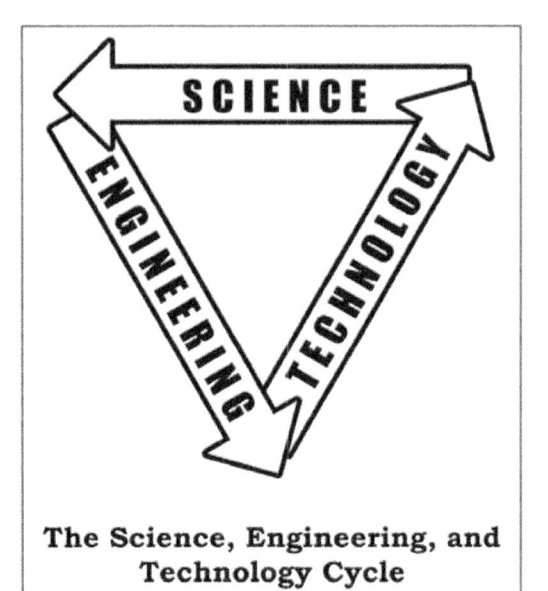

The Science, Engineering, and Technology Cycle

If an analogy were to be made to the religion of Christianity, **Science**, **Engineering**, and **Technology** could be viewed as a "**Holy Trinity**." **Science** is "**the Father**", the discipline that generates the theories and explanations about **Nature**. **Engineering** is the "**Blessed Mother**", the discipline which uses Science to bring forth the "**Holy Child**" which is **Technology**. In Christianity, the "**Holy Child**" is Jesus, and there are several interesting parallels between Jesus in the Bible to Technology. First, in the Bible, in the book of **Mark chapter 6 verse 3**, Jesus is called a "**Tekton**" which is the etymological root of the word "**Technology**." Moreover, several of the miracles associated with Jesus can be accomplished by way of Technology:

- It is said that Jesus can resurrect the dead, and technology in the form of Defibrillators can be used to "resurrect" someone if they experience heart failure.

- It is said that Jesus healed the sick, and technology in the form of Medication is used to heal the sick.

- It is said that Jesus walked on water, and technology can be used to facilitate the ability to walk on water.

- It is said that Jesus turned water into wine, and technology in the form of adding Phenolphthalein solution to alkaline water will make water appear as wine.

- It is said that Jesus can give you everlasting life, and technology in the form of "life support" can provide everlasting life

The holy books of the Abrahamic religions of Judaism, Christianity, and Islam are essentially tools, or technology that can be used and applied as needed. If it seems these religions promote an anti-Science theme then that is because the individuals using the tools are choosing to promote an anti-Science theme. One need only highlight the verses and instances in these books that promote science in order to use the tools to build a pro-science congregation. So not only is Jesus referred to as a tekton in Mark 6:3, and the word tekton means builder, craftsman, mason, or "technologist", but in order to be "Christ-like" you should want to be a tekton too. In the Talmud, Jesus is referred to as **Naggar bar naggar** meaning "Craftsman son of a Craftsman." Also, the Major Prophets from these holy books were Scientists. When **Noah** builds the ark, he too is acting as a **master builder**, craftsman, or "tekton." In the book of Genesis, after God creates, God has to look at his creation to "see that it is good," which means there is a possibility that it may not have been good, which means God was observing his experiment like a Scientist. When Adam names the animals in Genesis 2:20, he is essentially acting as a **Taxonomist**. When Eve has two competing hypothesis from God and the Serpent about whether or not she will die if she eats from the tree of knowledge, Eve performs the Science experiment and takes a bite of the fruit. When Abraham crosses over the Euphrates River into the land that would become Israel that means he had a hypothesis that he wanted to investigate. And in the religion of Islam, the titles of the chapters revealed to Muhammad like Al-Araf meaning "the Faculty of Discernment," Al-Furquan meaning "the standard criterion of truth and falsehood," Fussilat meaning "expounded, explained in detail, clearly spelled out," and Al-Bayyinah meaning "clear evidence and clear truth" refer to scientific concepts.

In the religion of Christianity, the "Blessed Mother" is not part of the "Holy Trinity," but rather the "Holy Trinity" is comprised of "the Father, the Son, and the Holy Ghost." The discipline of Mathematics can be likened to the "Holy Ghost" or "Holy Spirit" because like the "Holy Ghost," Mathematics is not tangible, it is an abstraction that cannot be seen, but rather you see the effects, results, and use of it. Also, in Christianity the "Holy Spirit" is associated with the "Word of God," and Mathematics is the "Word" or "language" of Science and the Universe. Every religion has its language. They say, if you cannot speak the language of Hebrew, then you cannot really understand the Bible; if you cannot speak the language of Arabic, then you cannot really understand the Quran and the religion of Islam;

if you cannot read the language of the MeduNeter, then you cannot really understand Ancient Kemet. Then it follows, if you do not know Mathematics, which is the language of Science and the Universe, then you cannot truly comprehend **Science**, **Engineering**, and **Technology**.

Instead of a "Holy Trinity," **Science**, **Technology**, **Engineering**, and **Mathematics** form a "**Holy Quad-ity**." The Quad-ity (4 in 1) is represented by the four letters: **S.T.E.M.** The four letters **S.T.E.M.** not only stand for **Science**, **Technology**, **Engineering**, and **Mathematics**, which are the disciplines and schools of thought dedicated to studying and understanding Nature, and harnessing, controlling, and working with the laws of Nature, but the four letters S.T.E.M. also stand for **Space**, **Time**, **Energy**, and **Matter** - everything there is in Nature and the Universe. In the science called Physics, which means nature, you learn about the **Space-Time** Continuum in Astrophysics on the Macrocosmic scale, and you learn about the **Matter-Energy** Continuum given by the equation $E=mc^2$ in Relativistic Physics or Quantum Physics on the Microcosmic scale. Everything there is in Nature and the Universe can be classified as **Space**, **Time**, **Energy**, or **Matter**. Thus, the letters **S.T.E.M.** stand for SPACE, TIME, ENERGY, and

The Symbolism & Significance of Spectroscopy to Science & Survival

MATTER, all of Nature and the Universe, represented by the **circle**, and The letters **S.T.E.M.** stand for SCIENCE, TECHNOLOGY, ENGINEERING, AND MATHEMATICS, the study, comprehension, and practical application of the study of nature, represented by the square.

In education there are efforts to emphasize "**Art**" on the same level as **Science**, **Technology**, **Engineering**, and **Mathematics** to create "**S.T.E.A.M.**" (Science, Technology, Engineering, Art, and Math) based curriculum. While it is important for educational curriculum to be well rounded and include many subjects, experiences, and opportunities for students to learn, Art is distinctively different from Science, Technology, Engineering, and Math when it comes to the critical thinking, logic, and mental analytic skills required. Even within the fields of Science, Technology, Engineering, and Math there is a hierarchy in terms of the level of logic and critical thinking skills needed for each field. If Science, Technology, Engineering, and Math were to be ranked from most logic and critical thinking needed to least logic and critical thinking needed, these fields would be ranked as Mathematics, Science, Engineering, and Technology respectively. This order also makes sense because each subsequent field relies on the other: Science depends of Mathematics, Engineering depends on Science, and Technology depends on Engineering. The fields of Science, Technology, Engineering, and Math utilize **convergent thinking** whereas **Art** utilizes **divergent thinking**. The difference between Science, Technology, Engineering, and Math versus Art is also illustrated by the "**Left Brain versus Right Brain**" paradigm discussed in psychology which implies that certain parts of your brain are dedicated to functions like logic, reason, problem solving and reality, whereas other parts of your brain are dedicated to functions like emotion, imagination, and fiction (although this dichotomy does not exclusively occur on the left and right side of your brain). Moreover, Science, Technology, Engineering, and Mathematics are the tools most needed for survival, well-being, self-determination, and liberation, and emphasizing Art on the same level as Science, Technology, Engineering, and Mathematics may make people think they can liberate themselves through artistic expressions like drawing a picture or singing a song. Just like the subjects of Reading and Writing, Art is currently used in all of the fields of Science, Technology, Engineering, and Math. Just like it is not necessary to emphasize Reading and Writing in S.T.E.M., it is not necessary to emphasize Art in S.T.E.M. because it is already present in every field. Art is used in every field in S.T.E.M. to express concepts and for **visualization**. Conversely, you can use Science, Technology, Engineering, and Mathematics to create and analyze Art. However, it is possible to make Art without using S.T.E.M. and it is possible to teach and learn S.T.E.M. without using Art, but it works best when both sides work together.

As I conclude this chapter on sharing my perspective, I would like to re-emphasize how important Science is to survival and well-being, and how important observation is to Science. The etymological origin of the word "**Science**" comes from words meaning "**knowledge and learning**." We are all students of life. Another word for "student" is "**pupil**," which is also the word used to describe **the central part of the eye which facilitates sight and observation**. This implies that as pupils, we acquire knowledge or Science through observation. Although I was raised in a religious tradition, the perspective that I share with you here is not based on religion. I come to you not as a Muslim or a Christian; not as a Hebrew or a Jew; not as a Buddhist or a Hindu; not as an Atheist or Agnostic. I do not come to you as a member of any religion, sect, group, cult, or organization. I simply come to you in the same way in which **we all were born into this world**, as a human being **endowed with senses to experience the world around us**, and a mind, that uses logic and reason to process and make sense of our experiences so that we can make the best choices in life, and plan the best course of action. And with that simple realization that this is the way we all come into this world, **I bear witness to the reality, that we are all born Scientists**. Science, at its root, at its core, at its essence, is nothing more than the systematic observation and study of nature for the purpose of gaining **knowledge, wisdom,** and **understanding**, and **making predictions**. We are all born into this world as scientists, and we were scientists before we were any of the things that we identify ourselves as. Before you belonged to any religion; you were a scientist. Before you knew you were Black; you were a scientist. Before you knew you were a boy or a girl, male or female; you were a scientist. Before you were Christian, Hebrew, Muslim, African, and/or Kemetic; you were a scientist! Why do I say that you were a scientist before any of these things? Because in order to identify with any of those things in the first place, you had to be a scientist first, you had to observe yourself, observe your surroundings, make a hypothesis, an educated guess, and come to a conclusion. That is the Scientific Method. You were all born scientists! Every step you take, every bite of food you eat, every breath you take, every decision you make, you are performing a scientific experiment; you are testing a hypothesis. Science is your way of life whether you realize it or not. So as I conclude, I would like to emphasize that Science,

Technology, Engineering, and Math, are the keys to survival, well-being, freedom, and liberation. As I previously mentioned, we are all born scientists, and we were scientist before we were anything that we identified ourselves with. It was and is through science that we were, and are able to survive. The way our ancestors knew which plants to eat and which plants not to eat was through science, trial-and-error; experimentation. The way our ancestors knew where to go, where to migrate to, which places to avoid, and which places in which to settle, was through science. You must comprehend that you are here today, your very existence, your very subsistence, your very LIFE, was made possible through science. You are the result of a long series of continuously successful scientific experiments carried out by your ancestors, going all the way back, from multi-cellular organisms, to single-celled organisms, to atoms, the elements, to the very fundamental particles of nature, like the electron, which dwells within each of us. The moment you abandon science, is the moment you start courting extinction. The moment you abdicate Science, you pursue your own elimination and annihilation. But the moment that you advocate for Science, the moment that you decide to start actively using and practicing science, when we have an enthusiastic passion for science, and a religious devotion and zeal for science, then we will see ourselves begin to truly thrive, prosper, and succeed. I can do all things through Science who strengthens me. So,

Let us use our Minds to discover

*the **Science** to accept the things we cannot change,*

***Technology** to **Engineer** the things we can,*

*and the **Mathematics** to know the difference.*

3. *Spec*ulation

The etymological origin of the word "**Speculation**" comes from words meaning "*observation and contemplation*", while the definition of the word "**Speculation**" refers to the concepts of "**assumption, hypothesizing, conjecture, faith, hope,** and **belief**." Hence, the word "**Speculation**" can be associated

Eye Question Mark

with both the concepts of "**Faith**" and "**Sight**," and thus, the verse from **2 Corinthians 5:7** of the **Bible** which states "*For we walk by faith, not by sight*" can be restated as simply "*We walk by Speculation!*" In this chapter I will discuss various symbols, icons, concepts, and mythologies from African **belief** systems, and other belief systems from around the world, related to **observation**. Looking backward in time, I **speculate** and **hypothesize** about if these symbols related to **observation** were indeed used and applied in a scientific context to their respective cultures. Looking at the present and looking forward in time into the future, I recommend that these symbols be used and applied in a scientific context in the science of **Spectroscopy**.

One question I often speculate about is, "*What came first, the Observation, or the Hypothesis?*" For example, if we could travel back to the beginning of time, and enter into the mind of the first conscious being, was an observation made about the universe which then led to a question or hypothesis? Or, did a question or hypothesis form in the mind, which in turn required an observation to answer the question? In either case, this would have been the first instance of Science. In Science, you can ask questions about existing data that you already have, or you may ask a question that requires you to have to gather new information and collect new data. The cyclic nature of Hypothesis and Observation is why the **Scientific Method** is best modeled by a **circular diagram**. New observations lead to new questions, which lead to new observations, and the cycle keeps going. I guess you can say, well if there have been no observations, then there would be no information, and thus no need to ask any questions, so the observation must have come first. However, the mere concept of observation or detection requires something to receive the information, which implies a "mind" of some sort wanting information for some reason (i.e. to answer some question). In the "**Chicken and the Egg paradox**", **evolution** teaches us that the **egg came before the chicken**, but the being that laid the egg that gave birth to the first chicken was not a chicken, but rather an early ancestor of the chicken. So using **analogical reasoning**, I speculate that the "Hypothesis" (Reason) came before the "Observation," but the mind of the being that asked the question which led to the first observation, was not the same being that made the first observation, but rather the creator of the being that made the first observation. While this may sound like "**Creationism**," I assure you it is not. If we think of "Observation" as **receiving light (or information)** and "Hypothesizing" as sending out or **emitting light (or requesting information)**, then these actions can be seen to metaphorically occur at the subatomic scale.

First we must comprehend **What is the Origin of Light**? Light, or **Electromagnetic Radiation**, is caused by the movement of Electrons within an Atom. In order to comprehend the origin and cause of Electromagnetic Radiation, we must first comprehend the structure of the **Atom**.

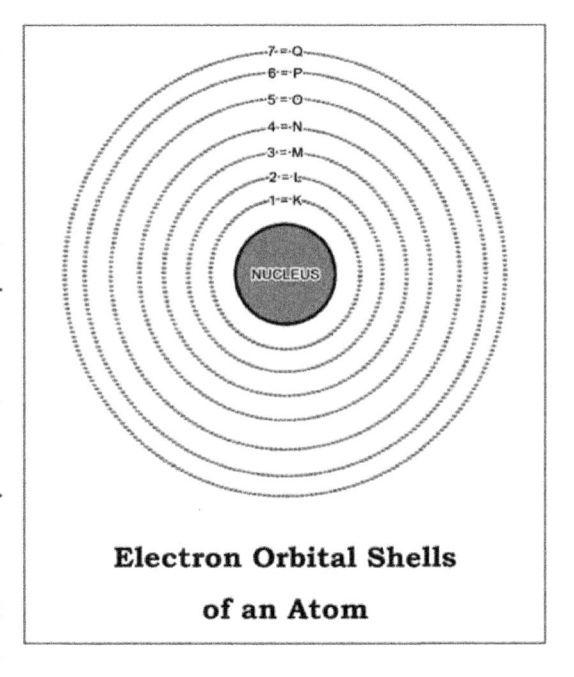

Electron Orbital Shells of an Atom

At the center of the Atom is the Nucleus, and around the nucleus of the Atom are seven (7) Orbital shells in which **Electrons** circle around the nucleus of the Atom. When an Atom **absorbs Energy** or "**Receives Light**", the Electrons move from the lower Orbital shells near the Nucleus to the higher Orbital Shells away from the Nucleus. When the Electrons move back to the lower Orbital shells near the Nucleus from the higher Orbital shells away from the Nucleus, a **Photon** (light particle) of Electromagnetic Radiation or **light is emitted from the Atom**. The movement of Electrons from each orbital shell in various Atoms produces the various frequencies and colors of the Electromagnetic Radiation Spectrum. The table below shows the predicted Electromagnetic Radiation emitted from an Atom of Hydrogen.

Electron Orbital Shell FROM	TO	Frequency (THz)	Wavelength (nm)	Electromagnetic Radiation Type
7	1	3297	91	ULTRAVIOLET
6	1	3191	94	
5	1	3158	95	
4	1	3093	97	
3	1	2913	103	
2	1	2459	122	
7	2	756	397	UVA - Black Light
6	2	732	410	VISIBLE - Violet
5	2	691	434	VISIBLE - Blue-Indigo
4	2	617	486	VISIBLE - Cyan (Blue-Green)
3	2	457	656	VISIBLE - Red
7	3	294	1020	INFRARED
6	3	274	1094	
5	3	234	1282	
4	3	160	1875	
7	4	138	2170	
6	4	114	2630	
5	4	74	4050	
7	5	64	4650	
6	5	40	7460	
7	6	24	12400	

The source and origin of Electromagnetic Radiation, or "Light" is the movement of electrons inside the orbital shells of an Atom. On Earth, the primary source of "Light" is the **Sun**. Thus, Sacred Geometry is observed in the fact that Macrocosmic **Solar Systems** resemble microcosmic **Atomic systems**. In fact, in the book **"Stolen Legacy"** by **George G.M. James**, it is suggested that the Greek word **"Atom,"** meaning *"uncut or indivisible,"* was derived from the Ancient Egyptian Sun deity **ATUM**, meaning *"complete"*.

The Symbolism & Significance of Spectroscopy to Science & Survival

Moreover, the Ancient Egyptian cosmological story which discusses the birth of **ATUM** parallels the scientific explanation for how light is emitted from an **Atom**. Cosmology is the science of the origin and development of the Universe. Scientific cosmology is the scholarly and scientific cosmology whereas Religious or Mythological Cosmology is a way of explaining cosmology based on religious mythology. NASA (The National Aeronautics and Space Administration) states that: *"Modern scientific cosmology is dominated by the **Big Bang Theory**, which **attempts to bring together observational astronomy and particle physics**. The Big Bang Theory is the prevailing model in modern cosmology that describes the* **origin and development of the Universe in the beginning of time, from an extremely hot and dense state** which began expanding rapidly. After the initial expansion, the Universe cooled sufficiently so energy could be converted into various **subatomic particles (quarks and electrons)**. After subatomic particles were formed, **Atoms and elements** began to form such as Hydrogen, Helium, and Lithium. Giant clouds of these primordial atomic elements later coalesced through gravity to

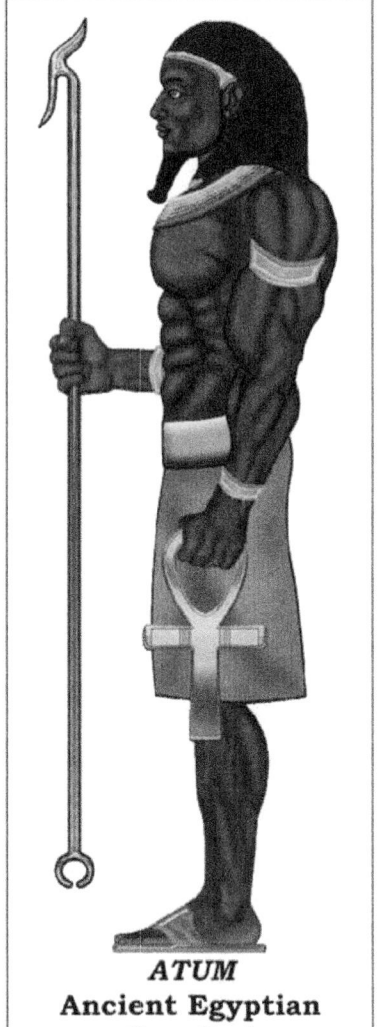

ATUM
Ancient Egyptian Sun God

*form **stars** and **galaxies**, and the heavier elements and planets were formed either within the stars, or during supernovae."*

In summary, the Big Bang Theory consists of 4 parts:

1. Primordial Abyss – the extremely hot and dense state
2. Sub-atomic particles (Quarks and Electrons) form
3. Atoms and light elements begin to form
4. Stars Galaxies and planets are formed

The Ancient Egyptian Memphite Theology contains the Theological, Cosmological, and Philosophical views of the Ancient Egyptians. The Memphite Theology was inscribed on an artifact called the "Shabaka Stone" by the Ancient Egyptian **25th Dynasty Nubian Pharaoh Shabaka Nefer Ka Re**. The Memphite Theology describes how **PTAH**, the craftsman deity, was fundamental in the creation and cosmology of the Universe. It is said that **PTAH** rises from the **NUN** (primordial abyss) as the primordial mound, and then creates **ATUM** in a younger (smaller) form named "**NEFER-Tum**", who then matures to become **ATUM-RE** (the sun) who

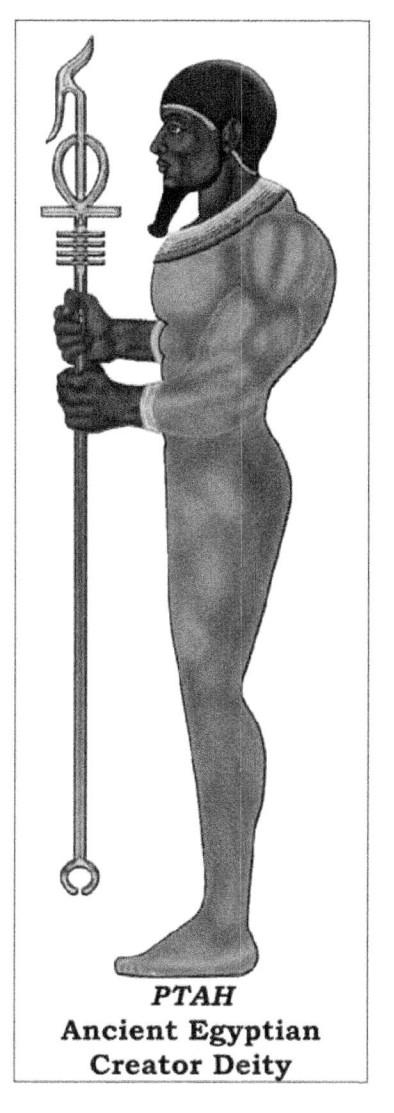

PTAH
Ancient Egyptian Creator Deity

The Symbolism & Significance of Spectroscopy to Science & Survival

creates the 8 other beings (**SHU**, **TEFNUT**, **GEB**, **NUT**, **ASAR**, **ASET**, **NEB-HET**, **SUTUKH**) who together form the Ennead. In the book entitled "***Stolen Legacy***" by George G.M. James, it is suggested that the very concepts of **logic** and **reason** ("*logos*" and "*nous*") in Greek philosophy were taken from concepts within the Ancient Egyptian Memphite Theology about a deity named **PTAH**. Also, the books "***Stolen Legacy***" and "**African Philosophy: The Pharaonic Period: 2780-330 BC**" by **Dr. Théophile Obenga** discuss how the **Atomist Philosophy** of the Ancient Greeks, which gave birth to modern **Atomic Theory**, was derived from concepts related to the Ancient Egyptian deity named **ATUM**. The **Memphite Theology** can be seen as a "**Spiritual Science**" in that it was part of the esoteric Religious Theology and Mythology of the Ancient Egyptians, and the Memphite Theology also shares many similarities with our modern scientific "Big Bang Theory" cosmology. To summarize, the cosmology in the Memphite Theology consists of 4 parts:

1. Primordial Abyss (**NUN**)
2. **PTAH** rising from the Primordial Abyss
3. **PTAH** creates **ATUM** (**NEFER-TUM**)
4. **ATUM** (**ATUM-RE**) creates the **ENNEAD**

As we can see, the four stages of the Big Bang Theory Cosmology align well with the Memphite Theology Cosmology. Based on this alignment, we can liken **ATUM** in his smaller and younger form as **NEFER-TUM** to the **Atom**, and **ATUM** in his matured form as **ATUM-RE** to the **Sun**, which is a **Star**.

Comparison of Big Bang Theory Cosmology to the Memphite Theology Cosmology

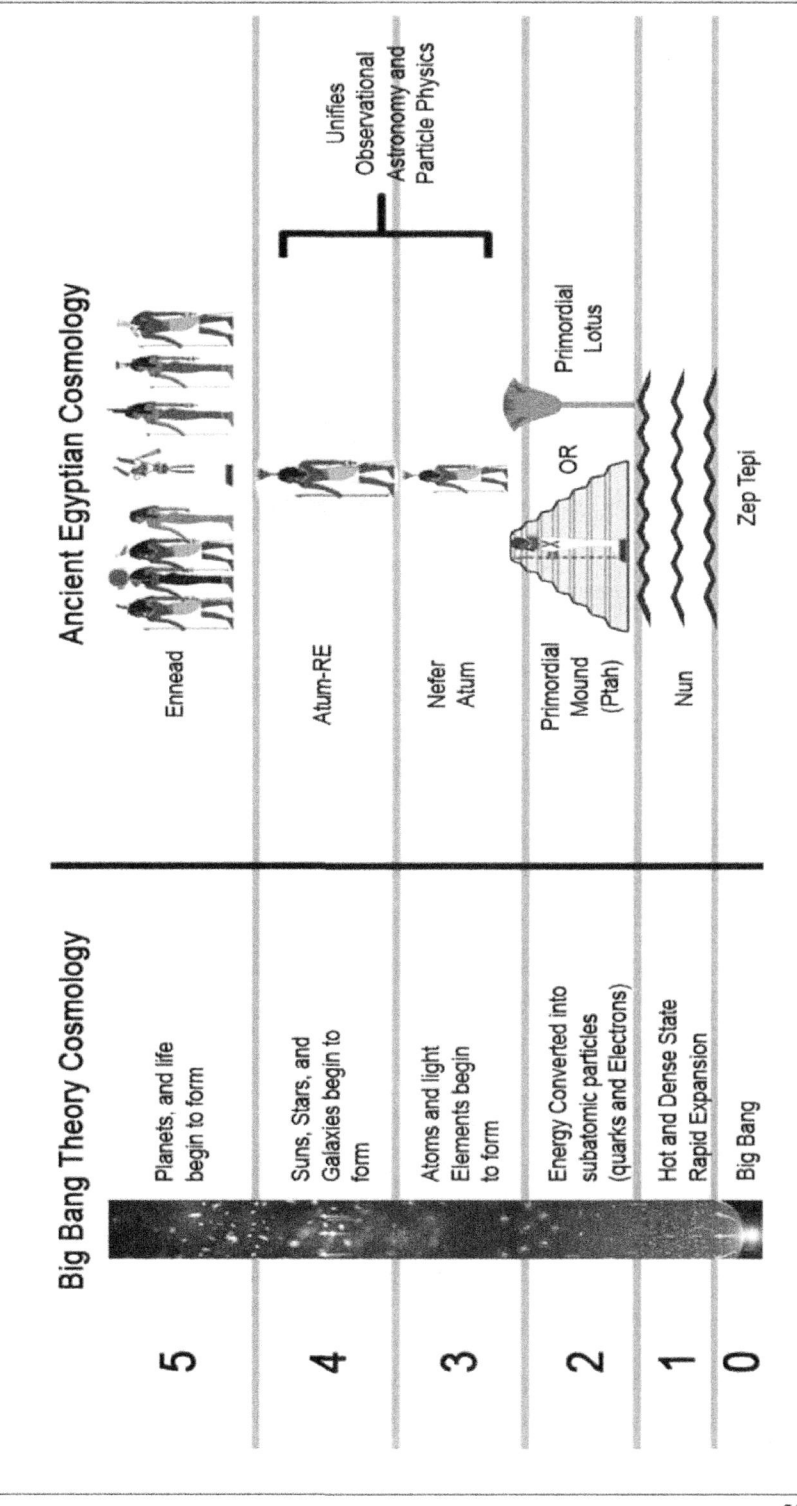

Big Bang Theory Cosmology

5 — Planets, and life begin to form

4 — Suns, Stars, and Galaxies begin to form

3 — Atoms and light Elements begin to form

2 — Energy Converted into subatomic particles (quarks and Electrons)

1 — Hot and Dense State Rapid Expansion

0 — Big Bang

Ancient Egyptian Cosmology

Ennead

Atum-RE

Nefer Atum

Primordial Mound (Ptah) OR Primordial Lotus

Nun — Zep Tepi

Unifies Observational Astronomy and Particle Physics

We can liken **PTAH** to the **Electron** sub-atomic (sub-**ATUM** or "below **ATUM**" as the primordial mound) particle. The function of **ATUM** in the **Memphite Theology** cosmology also unifies **Observational Astronomy** and **Particle Physics**. The 8 other beings who were part of the Ennead created by **ATUM-RE** in the **Memphite Theology** can be likened to the 8 planets created by the Sun. The actions of **PTAH** creating **ATUM** (**NEFER-TUM**) can be likened to Particle Physics in that the sub-atomic and atomic must be created prior to the creation of the Sun, Stars, Solar Systems, and Galaxies. Thus, **PTAH** rising from the primordial abyss to create **ATUM** can be seen as a representation of Particle Physics. It is the work of **PTAH** which produces **ATUM** (light), and it is the movement of **Electrons** which produce **light**. In Ancient Egyptian cosmology, **ATUM** was one of several aspects of the "Sun" deity named **RE**, also spelled **RA**. The sun deity **RE** had a variety of aspects, each of which was represented by a different symbol. The aspect of **RE** named **ATUM** was personified as a man. The sun deity **RE** was merged with **HERU** or **Horus** to become *RE-Horakahty* symbolizing the two horizons of "morning and evening". *RE-Horakahty* was symbolized by the **falcon** or hawk, or a man with the head of a falcon or hawk. In the

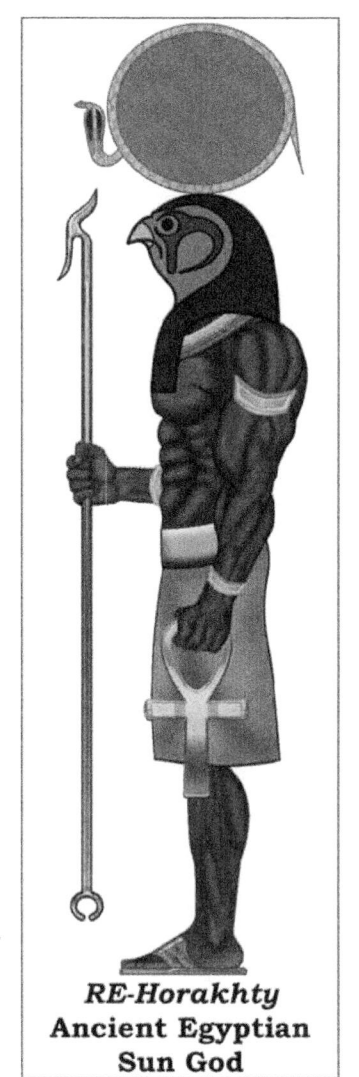

RE-Horakhty
Ancient Egyptian Sun God

Pyramid texts and Ancient Egyptian cities of Hermopolis and Heliopolis, **PTAH** is replaced by a **Lotus flower** in the cosmology. In the "**Lotus Flower**" version of the Theology, it is the **Lotus Flower** which rises from the primordial abyss. After rising from the primordial abyss, the **Lotus Flower** opens and it **gives birth to Light** in the form of **ATUM**, who goes on to create the eight other beings of the Ennead. Other variants of the cosmology depict different forms of Solar Deities such as **RE** or **HERU** emerging from the Lotus Flower as Light.

One version of this cosmology is depicted in the images known as the "**Dendera Lights**" from Ancient Egypt. The "**Dendera Lights**" refer to **7** unique images

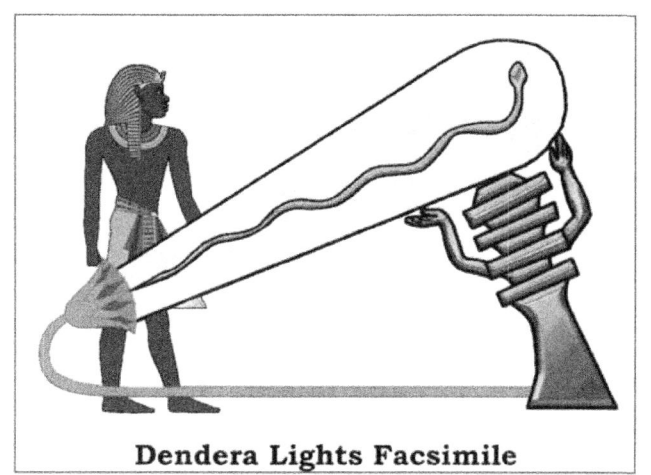

Dendera Lights Facsimile

found in two buildings from the **Hathor Dendera Temple Complex** in Ancient Egypt which depict objects which look similar to modern **electric light bulbs**; although it is most likely that these objects were just statues. The images associated with the "**Dendera Lights**" actually depict a "**Lotus Flower**" version of the aforementioned **Memphite Theology Cosmology** with the character of **ATUM** replaced with a version of **RE** and **HERU** called **HAR-SEMA-TAWY** (meaning "*HERU the unifier*"). The character of **HAR-SEMA-TAWY** has two forms, the primordial Serpent emerging from the Lotus Flower and the solar Falcon, and thus unifies (**Sema-Tawy**) the concepts of **Observational Astronomy** and **Particle Physics** much like the character of **ATUM**.

55

The Symbolism & Significance of Spectroscopy to Science & Survival

Comparison of Big Bang Theory to Ancient Egyptian Har-Sema-Tawy Atomist Philosophy

Level	Big Bang Theory Cosmology	Ancient Egyptian Cosmology
5	Planets, and life begin to form	Ennead
4	Suns, Stars, and Galaxies begin to form	Har-Sema-Tawy (Falcon)
3	Atoms and light Elements begin to form	Har-Sema-Tawy (Primordial Snake)
2	Energy Converted into subatomic particles (quarks and Electrons)	Primordial Lotus
1	Hot and Dense State Rapid Expansion	Nun
0	Big Bang	Zep Tepi

Har-Sema-Tawy Unifies Observational Astronomy and Particle Physics

The Ancient Egyptians would build Pyramids and temples such that rays of starlight or sunlight would enter into the structure on certain significant dates and times of the year. Once such temple is the **Great Temple of Ramses II** built in the village of **Abu Simbel** in the "**Nubian Monuments**" near the border of the modern day countries of **Egypt** and **Sudan**. The outside of the temple has four colossal statues of **Ramses II** with a statue of *RE-Horakahty* over the entrance of the temple. On the back wall of the temple are four seated statues of **PTAH**, **AMUN-RA**, **Ramses II**, and *RE-Horakahty*. The statues on the back wall of the temple remain in darkness for most of the year. However, the temple was designed such that twice a year, on February 22 and October 22, the date of Ramses II's birthday and coronation, a ray of sunlight is able to enter the temple and be **dispersed** to **illuminate** the statues of **AMUN-RA**, **Ramses II**, and *RE-Horakahty* while the statue of PTAH remains in darkness year round. The way in which the **Great Temple of Ramses II** was designed and built to capture and disperse a ray of light is similar to the way Spectrometers work as will be discussed in the following chapters.

Drawing of statues of PTAH, AMUN-RA, Ramses II, and *RE-Horakahty* from Great Temple of Ramses II at Abu Simbel

The analogies of the "**Electron and Photon**" to "**PTAH and ATUM**" to "**Hypothesizing and Observation**" respectively, illustrate the point made earlier that "Hypothesis came before Observation, but the first instance of each was not within the same being". **PTAH** ("reason") represents "the **Electron**," and it is the action of the **Electron** which leads to "**emitting light**" or "**Hypothesizing**," and the emission or creation of light represents **ATUM**. Once light is emitted, it can be received or "**observed**," which may be the reason why many of the African **Eye** symbols related to **observation** are associated with natural phenomenon which emit light like the **sun** and the **moon**.

Eye of HERU (Horus)/ Ey of RA	Eye of DJEHUTI (Thoth)

In Ancient Egypt, the "**Eye of HERU (Horus)**," also known as the "**Eye of RA**" or **WADJET**, was one symbol related to observation which represented the **Sun**. **WADJET** was represented by the symbol of a **Cobra**, and appeared as the **Uraeus** snake in the center of Ancient Egyptian crowns. While the **Right Eye** symbol was associated with the **Sun**, the **Left Eye** symbol was associated with the **Moon**, and the lunar deity **DJEHUTI (Thoth)**. In Science, the word "**observation**" does not just refer to information received through sight, but rather all information capable of being received through our senses either naturally, or aided with technology. One interesting correlation between the "**Eye of HERU**" observation symbol from Ancient Egypt, to the concept of "**Scientific Observation**" is that the "**Eye of HERU**" symbol was used as a composite symbol to represent all of the information that can be received to the human mind by way of the senses (i.e. Scientific Observation). The table below describes what each of the component parts of the "**Eye of HERU**" symbol represented:

Symbol	Description	Sense
	Nose	Smell
	Eye Pupil	Sight
	Eyebrow	Thought
	Ear	Hearing
	Tongue	Taste
	Hand	Touch
	African Symbol for All Information received via the senses by The Human Mind	

The Symbolism & Significance of Spectroscopy to Science & Survival

The etymology of the word "**Horoscope**" means "*observation of the heavens*," and is phonetically similar to "*Horus Scopes*," and it is indeed the "**Eye of Horus**" which "**scopes**" or **observes** the heavens when a Falcon or Hawk is flying. Another aspect of the Sun deity RE was the sun disc called ***ATUN*** or ***ATON*** or ***ATEN***. The **ATON** sun disc represented a synthesis of all the aspects of **RE** in one symbol. The **ATON** was depicted as a sun disc or a **winged-sun disc** with various arms or "Rays" emerging from the disc. It is interesting to note that the word "*RE*" or "*RA*" is phonetically similar to the word "Ray" and can also be found in the word "Radiation". Another interesting observation is that the sun disc Aton is occasionally depicted with 14 Rays emerging from it represented the "**14 Creative powers of RE**".

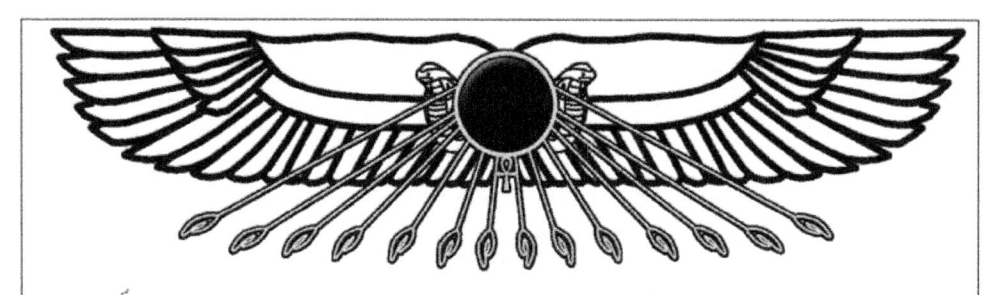

Aton or Aten the Sun Disc
With 14 "Rays" of Electromagnetic Radiation
as depicted by the Pharaoh Akhenaten

In modern science we also find the prevalence of the number 7 and 14 when we study phenomena related to Electromagnetic Radiation. As mentioned earlier, there are 7 Electron Orbital shells. There are also seven frequency categories of the Electromagnetic Spectrum, and seven color frequency categories of the Visible Light portion of the electromagnetic spectrum. When we add the 7 Electromagnetic Radiation spectrum frequency categories with the 7 color frequency categories of Visible Light we get the number 14. We can speculate about if this is what was symbolically depicted by the Ancient Egyptians in the 14 rays emerging from the *Aton* sun disc and the 14 "creative powers" of *RE*. However, for present and future practical application, the depiction of the *Aton* sun disc and the 14 "creative powers" of *RE* serves as an excellent **mnemonic device** to remember the various frequencies of the Electromagnetic Radiation Spectrum. The 7 Frequency Categories of the Electromagnetic Radiation Spectrum are:

TYPE	Frequency	Wavelength
Gamma ray	100,000,000 THz +	0nm TO 0.003 nm
X-Ray	30,000 THz TO 30,000,000 THz	0.01 nm TO 10 nm
Ultraviolet	800 THz TO 30,000 THz	10 nm TO 375 nm
Visible	400 THz TO 800 THz	380 nm TO 750 nm
Infrared	0.3 THz TO 400 THz	750 nm TO 1,000,000 nm
Microwave	0.0003 THz TO 0.3 THz	1,000,000 nm TO 1,000,000,000 nm
Radio	0 THz TO 0.0003 THz	1,000,000,000 nm +

The 7 Color Frequency Categories of the Visible Light Portion of the Electromagnetic Radiation Spectrum are:

Color	Frequency	Wavelength
violet	668–789 THz	380–450 nm
indigo	631–668 THz	450–475 nm
blue	606–630 THz	476–495 nm
green	526–606 THz	495–570 nm
yellow	508–526 THz	570–590 nm
orange	484–508 THz	590–620 nm
red	400–484 THz	620–750 nm

The Symbolism & Significance of Spectroscopy to Science & Survival

The frequencies between 400 THz to 800 THz of the Electromagnetic radiation spectrum are the frequencies that Human beings can see with our **eyes** and hence this portion of the Electromagnetic radiation spectrum is called the "**Visible Light Spectrum**". When all of the color frequencies of the Visible Light spectrum are combined, the product is called "**White Light**". There is symbolism within **West African Vodun** mythology which expresses the relationship between "**White Light**" and the "color frequencies" which make up the **Rainbow**. In **West African Vodun** (and also **Haitian Vodou** and **American Voodoo**), **Damballah** is the name of the <u>White</u> serpent who is considered the **sky deity**. The consort to the **White serpent sky deity Damballah** is named **Ayida Weddo** who represents the **Rainbow** and is called the "**rainbow serpent**" or the "**rainbow snake**."

This African mythology about the "**White snake**" being married to the "**Rainbow snake**" illustrates the relationship between "**White**

Veve depicting White Snake Damballah and the Rainbow snake Ayida-Weddo

Light" and the color frequencies which make up the **Rainbow**. In the **Yoruba** tradition, the "**rainbow snake**" is the **Orisha** named **Oshunmare**. The "**rainbow snake**" is also a part of the mythology of the indigenous **Aboriginal** people of Australia who regard the "**rainbow snake**" as the creator deity and created one of the earliest known depictions of a rainbow which dates back to around 4000 B.C.

Certain objects can be "**Heated**" or **Thermally Radiated** to the point where they emit Visible Light or "**White Light**". The emission of Visible Electromagnetic Radiation due to Thermally Radiating or Heating an object which dramatically increases its temperature is called **Incandescence**. Examples of Visible Light or "White Light" due to Incandescent Thermal Radiation are the **Sun**, **Incandescent Light Bulbs**, and **Fire**. In modern science, the name for an **ideal absorber** and **perfect emitter** of "White Light" due to Thermal Radiation is called a **Black-Body** and the Incandescent electromagnetic radiation emitted by a Black-Body is called **Black-Body Radiation**.

It is important to note the difference between the **color of light** and the **color of an object**. Colored light, like the light emitted from a **LASER**, is emitted due to the movement of electrons between orbital shells. Because the various color frequencies of visible light combine to produce "White Light", a "**White Light LASER**" requires the combination of multiple LASER sources of each color frequency. Combining or adding all the color frequencies of visible light to produce White Light is the principle behind the concept of **Additive Color**. The different color frequencies of several LASERs can be refracted and combined to form a "White Light LASER" using a **Prism**, and also "White Light" that is emitted from an Incandescent source like the Sun can be **refracted** and separated into each color frequency using a **Prism**. An object that does not emit light, but is perceived as having a color, is actually absorbing all of

The Symbolism & Significance of Spectroscopy to Science & Survival

the color frequencies contained within white light, and reflecting the one color that you see. For example, if you see a red apple, the reason why you see the color red is because the apple is absorbing all of the other color frequencies contained in white light (orange, yellow, green, blue, indigo, and violet) and reflecting the color frequency of red. When you see the color Green in plants, that means the plant is absorbing all the other color frequencies contained in white light (red, orange, yellow, blue, indigo, and violet) and reflecting the color frequency of Green. When you see a White colored object, that means the object is reflecting all of the color frequencies contained within White light and absorbing nothing. When you see that an object is the color **Black**, that means that the object is absorbing ALL of the color frequencies within visible light and is reflecting nothing. Absorbing all of the color frequencies of visible light to produce Black is the principle behind the concept of **Subtractive Color**.

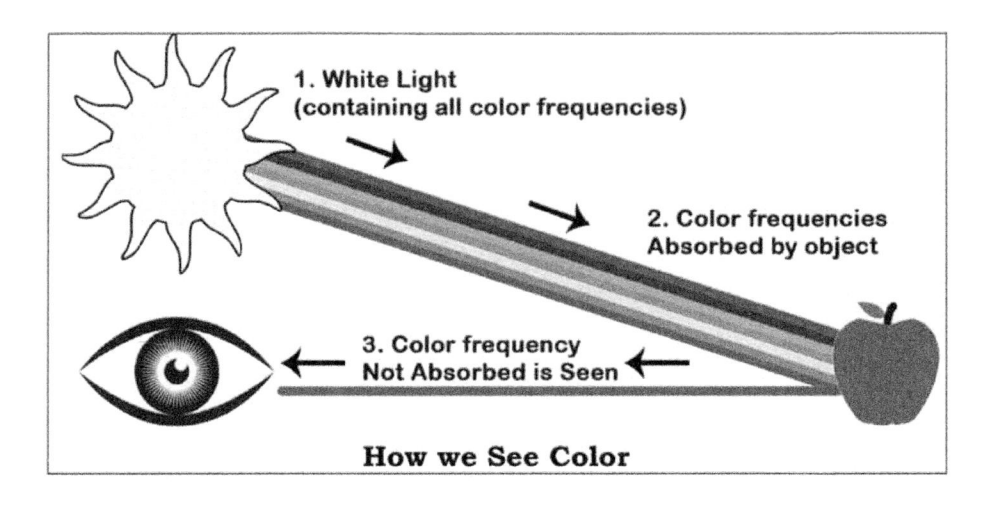

How we See Color

1. White Light (containing all color frequencies)
2. Color frequencies Absorbed by object
3. Color frequency Not Absorbed is Seen

In Physics, the idea that a **Black-Body** is both an **ideal absorber** and **perfect emitter** of "White Light" can be represented by the Ancient Egyptian deity **AUSAR (Osiris)** who is called "**Lord of the Perfect Black**" by **Dr. Frances Cress Welsing** in her book entitled "***The Isis Papers***." AUSAR is also called "**Kem-ur**" (**Kem-wer**) meaning "***The Great Black One***" several times in the "**Book of Coming Forth by Day**" commonly known as the "**Egyptian Book of the Dead**." In fact, the **Medu Neter** Hieroglyphics for the name of the Ancient Egyptian deity **AUSAR** contains an **Eye** symbol. In practical application, and everyday life, most of the "**Black**" colors that you see are not "**Perfect Black**." **Black 2.0** is the name of commercially available Black paint which absorbs 95% of visible light, and **Black 3.0** is the name of another more expensive commercially available Black paint which

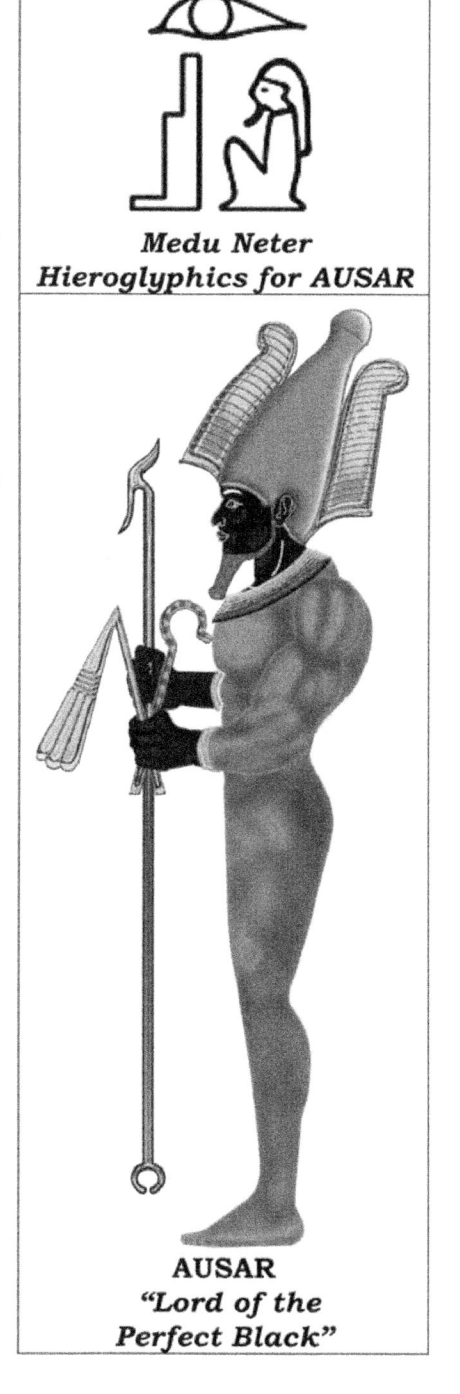

Medu Neter Hieroglyphics for AUSAR

AUSAR
"Lord of the Perfect Black"

absorbs 99% of visible light. **Super Black** is a substance developed by **National Physical Laboratory** which absorbs 99.6% of visible light. **Vantablack** is a substance that absorbs 99.965% of visible light which was developed in the year 2014 by the company **Surrey NanoSystems Limited.** The term "**Vanta**" in **Vantablack** stands for "**V**ertically **A**ligned Carbon **N**anotube **A**rrays". The vertical carbon nanotubes within the Vantablack substance essentially acts as a prison for light, trapping light within the substance by causing the light to continuously deflect and be absorbed by the carbon nanotube "jail bars" within the substance. To date, the "**Blackest Black**" is another material made from **carbon nanotubes (CNT)** which absorbs 99.995% of visible light and was developed by Engineers from the **Massachusetts Institute of Technology (MIT)** named **Brian Wardle** and **Kehang Cui**.

However, all of these "Blacks" are not *"the Perfect Black"*. In **Physics**, "**the Perfect Black**" **(AUSAR)** is a **Black-Body** which absorbs **100%** of visible light. It is important to reemphasize, that a **Black-Body** not only is a perfect absorber of light, but it is also a perfect emitter of light. **Emission of light** is different from **reflection of light**. Emission of light means the object is a "**Source of Light**". Reflection of light means that the object does not absorb light, but rather light "bounces" off of the object. In Physics, a "**White Body**" is an object that is a perfect reflector of light. In practical application, **Spectralon** is a material developed by the company named **Labsphere** in 1986 and is called the world's "**Whitest White**" which reflects 99% of visible light.

Until now, all **Eyes have been on Egypt**, and all the Eye symbols discussed so far related to observation have been from Ancient Egypt. However, other Eye symbols related to observation can be found throughout Africa and the world.

Adinkra Eye Symbols from West Africa	
Onyakopon Aniwa Symbol of the Omnipresence of God God's eye sees all secrets	
Abode Santann The all seeing eye of the Divine Creator Symbol of Universal totality	**Kojo Baiden** Rays Cosmos, Omnipresence "Moon"

In the West African Adinkra system of hieroglyphic symbols, there are three prominent Eye symbols. **Onyakopon Aniwa** is a symbol of the Omnipresence of God and represents the idea that "**God's Eye sees all secrets**." The **unification** of Human and Natural creation in the **totality of the universe** is expressed in the West African Adinkra symbol called "***Abode Santann***".

The Abode Santann symbol depicts 3 forms of creation in one symbol: The **Sun**, the **Moon**, and the Stool (Human or **Earth**). Much like the "**Eye of DJEHUTI**" is a horizontal reflection of the "**Eye of HERU**" in the Ancient Egyptian Hieroglyphic set of symbols, the ***Kojo Baiden*** eye symbol is a vertical reflection of the ***Abode Santann*** eye symbol in the West African Adinkra Hieroglyphic symbols. The ***Kojo Baiden*** eye symbol represents the Cosmos and divine Omnipresence. While the ***Kojo Baiden*** eye symbol is said to represent the Moon, the ***Abode Santann*** eye symbol is said to represent the sun. The sun was also personified in other African traditions as ***Anyanwu*** who represented the "eye of the Sun" in the **Igbo** culture and also as ***Lisa***, the sun deity in **Dahomey** culture, who was the sibling to the moon deity ***Mawu***, and offspring of the Orisha ***Nana Buluku***.

In Southern Africa, the greeting "**Sawubona**" means "*I see you*" and "**I respect you**". The response to the greeting can be "**Sikhona**" which means "*I am here to be seen*" or "**Shiboka**" which means "*I exist for you*". The triad of terms "**Sawubona, Sikhona, Shiboka**" meaning, "*I see you, I am here to be seen, and because you see me, I exist*" represents the important scientific principle learned from the "**Double-slit experiment**" in **Quantum Physics** that "**reality does not exist until it is observed**." This is also conveyed somewhat in the expression that "**Perception is Reality**" or "**What you see is what you get**" (WYSIWYG).

Eye symbols related to observation can be found in other cultures around the world. In **Buddhism**, the "**Eyes of Buddha**" represent Wisdom and Enlightenment, and symbolizes the vast

Eyes of Buddha

knowledge and **omniscience** obtained through being **all-seeing**, able to **observe** everything.

In the **Quran**, the Holy Book in the religion of **Islam**, in chapter 3 verse 190 it states: "***Behold!*** *In creation of the heavens and the earth, and the alteration of night and day,* ***there are indeed signs for men of understanding.***" This verse expresses the idea that knowledge is gained through observation (beholding). In **Islam**, the

Hamsa

27th attribute of the deity **Allah** is "Al-Basir" meaning "**the all seeing**." In the religion of **Islam**, and in the **Jewish** religion, the symbol called the **Hamsa** (also called the "**Hand of Fatimah**" or the "**Hand of Mary**") is a symbol of a hand with an eye in the center of the palm. The **Hamsa** symbol is used as a symbol of protection against bad luck, misfortune, and the "*evil eye*".

In **Hinduism**, the **third-eye** chakra is called **Ajna**, and is said to represent the "**Mind's Eye**" and the **Pineal gland** in the Brain. Some esoteric belief systems speculate that the "**Mind's Eye**" can utilize a sixth sense of **Extrasensory perception** (**ESP**) to perform acts such as **Clairvoyance** and **Remote Viewing**, which are the ability to see things, "**sight unseen**," in the past, present, or future without being physically present. "**Remote Sensing**" is the exoteric, operative, and scientific application of **Spectroscopy** to achieve the same objectives claimed by the esoteric speculative concept of "**Remote Viewing**." "**Remote Sensing**" through **Spectroscopy** facilitates the ability for Scientists to acquire information about distant stars and planets by analyzing light.

Ajna – Third-Eye Charka symbol

The "**God's Eye**" symbol within Native American culture represents concepts similar to **Remote Viewing** and **Clairvoyance**. The "**God's Eye**" symbol, called **Ojo de Dios** in Spanish, is believed to provide spiritual **insight** beyond normal physical sight. Another **Native American** eye symbol related to observation is the Mesoamerican **Aztec Calendar** called **Nahui Ollin**, meaning "Four Movements," which features an **eye symbol** in the center of the calendar graphic.

Native American God's Eye

Eye in the Center of the Aztec Nahui Ollin

Considering that the purpose of a calendar is to track the motion of the Earth around the Sun, then the Eye symbol at the center of the Aztec calendar is thought to be symbolic of the Sun. This has associations with the Sun being symbolized as an eye in Ancient Egypt, and also with the **Dikenga Cosmogram** from the **Kongo** called the "*Four Movements of the Sun.*"

Eye symbolism related to observation as a representation of concepts associated with knowledge, wisdom, and enlightenment are also found "**hidden in plain sight**" amongst modern day occult organizations, mystic traditions, and secret societies such as the

All Seeing Eye

Gnostics, **Sufis**, **Theosophists**, **Rosicrucians**, and **Freemasons**, just to name a few. One notable symbol is the "**All Seeing Eye**" or the "**Eye of Providence**" featured on the back of the U.S.A. dollar bill. One particular secret society that developed the motif of the entire organization primarily on principles centered on "eye symbolism" was the "**Great**

Enlightened Society of Oculist" established in the 1700s in Germany. An **"Oculist"** is an old term for what is now called an **Ophthalmologist** or **"Eye Doctor."** Much like the Freemason secret society operates as **speculative** masons utilizing symbolism related to operative masonry, the **"Great Enlightened Society of Oculist"** operated as metaphorical or **"speculative"** eye doctors, utilizing symbolism related to operative ophthalmology. The initiation ritual of the **"Great Enlightened Society of Oculist"** involved a metaphorical mock **"Eye Surgery"** where the **Oculist "Eye Doctors"** are said to **give sight to the blind**, so that the initiate can **see the light** and perceive the truth in reality. Examples of eye symbols related to observation and the practical application of **Spectroscopy** can not only be found in Religious mythology, but also in modern mythology in the form of Science Fiction. The concepts of Clairvoyance and Remote Viewing appear in the animated Television cartoon series called **Thundercats,** as the **Sword of Omens** giving **"Sight beyond Sight."** In the Science Fiction television series called **"Star Trek: The Next Generation,"** a blind character named **Geordi La Forge**, who is portrayed by actor **LeVar Burton**, utilizes a technological device called a **V.I.S.O.R. (Visual Instrument and Sight Organ Replacement)** which gives him the ability to see. On the television show, the **V.I.S.O.R.** technology is able to scan the entire **spectrum of electromagnetic radiation** and transmit sensory input to the brain, providing the wearer with more enhanced visual abilities than natural eyesight. It is interesting to note that the actor **LeVar Burton** also was the star on another television show called **"Reading Rainbow,"** and it is indeed the science of **Spectroscopy** which facilitates the ability to separate light into its component colors in order to "**Read the Rainbow**." From doors which open automatically based on motion detection, to Star Trek communicators foreshadowing cell-phones, there have been several forms of technology featured on the "**Star Trek**" Science Fiction television series which predicted the eventual development of a similar form of technology in real life. However, the science fiction technology from "**Star Trek**" which best demonstrates the enormous practical application of the science of **Spectroscopy** is the **Tricorder**. On "**Star Trek**" the **Tricorder**, or "Tri-function Recorder" (sensing, computing, recording), is a small device that can be held in one's hand and used to scan and analyze an object or the surroundings. On the television series, the **Tricorder** is used to inspect material objects, assess the atmosphere and terrain of unfamiliar environments, and examine and diagnose biological organisms. Since the **Tricorder** on the **Star Trek** science-fiction series functions by **scanning** substances, then if the object were engineered in real-life, it would function based on the principles of **Spectroscopy** and the interaction of electromagnetic radiation and matter. In recent years, there have been several technologies developed in real-life which mimic the function of the science fiction Tricorder. In 2005, scientists from **Purdue University** developed a compact **mass Spectrometer** based on **Desorption Electrospray Ionization (DESI)** which can scan and analyze materials on the go in the surrounding environment. In 2016, a company named **Consumer Physics** began developing a **molecular scanner** called **SCiO** which can interface with **smart-phones**.

The **SCiO** technology uses **spectrometry** to shine light on objects in order to analyze the chemical composition of the object. The intended use of the **SCiO** technology is to allow users to be able to scan food to determine ingredients and freshness, scan medicine to determine authenticity, scan surfaces to determine if any harmful toxin or virus is present, or scan your body to determine your body composition. In the coming years with the planned **"Internet of things (IoT)"** where all physical objects are planned to be fitted with sensors and software and networked through the internet, it will be the application of the science of **Spectroscopy** which will provide the data and information on the component parts of everything.

Depiction of person using a portable Spectrometer to obtain information about the environment in real-time

4. *Spec*troscopy

Spectroscopy is the scientific study of how matter interacts with electromagnetic radiation. The word **"Spectroscopy"** comes from the root words **"spec"** and **"scope"** both meaning **"to observe."** Since the way in which we are able to **observe** and see things is based on the interaction of light

Eye in Scope

with matter, then comprehending how we are able to see light and color serves as a good introductory starting point to discuss the science of **Spectroscopy**. **Electromagnetic Radiation** or "**Light**" is so fundamental to **Science** because the **speed of light** is a fixed fundamental constant of **Nature**, and it is used to define the base units used to measure **Space**, **Matter**, and **Time**, which are the **Meter**, the **Kilogram**, and the **Second**, respectively. **Electromagnetic Radiation** or "**Light**" can interact with **matter** in four different ways: 1) light can be **emitted** from matter, 2) light can be **transmitted** through matter, 3) light can be **absorbed** by matter, and 4) light can be **reflected** by matter. Light is emitted from matter when sufficient energy is applied to the matter to get it to emit light. Light is transmitted through matter if the matter is transparent or translucent and allows the light to pass through the matter. When light is transmitted through matter, the light may bend or be spread out through **dispersion** and/or

diffraction. Light is absorbed and/or reflected by matter after light from a source strikes the matter. Light absorbed by matter also causes a temperature change within the matter.

As mentioned in the previous chapter, the way we see color **emitted** from a light source like a **LASER** or **Neon Light** works different from the way we see the color **reflected** off of an object that does not emit light. Each color of light corresponds to a specific **wavelength** and **frequency**. The **color of a light emitted** by a light source occurs from the movement or "**excitation**" of electrons between orbital shells within the atom creating a "color of light" from a frequency or a combination of frequencies. The **color of an object** occurs when light from a source strikes an object, and the object **absorbs** some frequencies of the light, and **reflects** other frequencies of the light. The frequencies that are not absorbed by the object determine the "color" that you see. For example, if you see a "green plant", that means the plant absorbs all of the color frequencies of the visible electromagnetic radiation spectrum except for the green frequency. Conversely, if you see a "Green LASER" that means the LASER emits only the green frequency of the visible electromagnetic radiation spectrum. But the interesting question that gets into the science of Spectroscopy is, "*What makes certain objects reflect certain colors, and what makes certain light sources emit certain colors?*" The colors that we see, that is the frequency of light or electromagnetic radiation that is reflected from an object to our eyes, or emitted from a light source, is due to the atomic and molecular structure of the object or light source. For example, most atomic elements are gray or silver in color, but there are certain atomic elements that have a unique color. The table below shows the color of certain atomic elements.

SPEC-RA-SCOPE

Number	Symbol	Element Name	Color
5	B	Boron	Black
6	C	Carbon	Black
9	F	Fluorine	Yellow
16	S	Sulfur	Yellow
17	Cl	Chlorine	Yellow/Green
29	Cu	Copper	Brown
30	Zn	Zinc	Blue/White
35	Br	Bromine	Red
48	Cd	Cadmium	Blue/White
53	I	Iodine	Violet/purple
55	Cs	Cesium	Yellow/white
76	Ox	Osmium	Blue/White
79	Au	Gold	Yellow

The "color" or frequency of electromagnetic radiation reflected by each atomic element is determined by the frequency or movement of the electrons within the atom. In **Atomic Spectroscopy**, the wavelengths and frequencies of electromagnetic radiation absorbed by the atomic element is called the **Atomic Absorption Spectrum**. Conversely, if energy is applied to the atomic element to excite the electrons of the element and get the element to emit light, then the wavelengths and frequencies of electromagnetic radiation emitted by the atomic element is called the **Atomic Emission Spectrum**. While generating an **Atomic Absorption Spectrum** generally preserves the object being analyzed, generating an **Atomic Emission Spectrum** may **destroy** the object being analyzed because in order to get the object to emit light, the electrons of the object have to be "excited" causing a chemical change in the object.

Atomic Emission Spectra can be generated by either burning the atomic element in a flame or electrifying the element. For example, the element Lithium (Li) emits a red colored flame, Calcium (Ca) emits an orange colored flame, Sodium (Na) emits a yellow colored flame, Boron (B) emits a green colored flame, Cesium (Cs) emits a blue colored flame, and Potassium (K) emits a violet colored flame. When the **Noble Gas** elements are electrified, the element Helium (He) emits a pink colored light, Neon (Ne) emits a red colored light, Argon (Ar) emits an indigo colored light, Krypton (Kr) emits a green colored light, and Xenon (Xe) emits a blue colored light. When an **Atomic Emission Spectrum** is generated from a solid atomic element it is also called a **Thermal Spectrum** (due to the object heating up) and a **Continuous Spectrum** (due to the object emitting energy at all frequencies). When an **Atomic Emission Spectrum** is generated from a Gas, it is called a **Discrete Spectrum** (due to the gas emitting energy only at certain frequencies). The **Absorption Spectrum** and the **Emission Spectrum** for a given atomic element are mirror images of each other. Each atomic element has its own unique and different **Spectral lines** (which look similar to a "bar code") which serve as a signature or fingerprint that can be used to identify an element just by using light. However, the matter in the universe is not just composed of stand-alone atomic elements, but rather combinations of atomic elements called **molecules**, and electromagnetic radiation is a much broader spectrum of frequencies than just the colors of the visible light spectrum, but the principles discussed so far help to comprehend the basic principles of **Spectroscopy**.

SPEC-RA-SCOPE

Schematic of Creation of Emission and Absorption Spectral Lines

Emission Spectra

Excited Matter Emits Light — Slit — Prism or Diffraction Gradient — Dispersed Light — Detector — Emission Spectral Lines (Hydrogen Example)

Absorption Spectra

White Light Emitted — Matter Absorbs Light — Slit — Prism or Diffraction Gradient — Dispersed Light — Detector — Absorption Spectral Lines (Hydrogen Example)

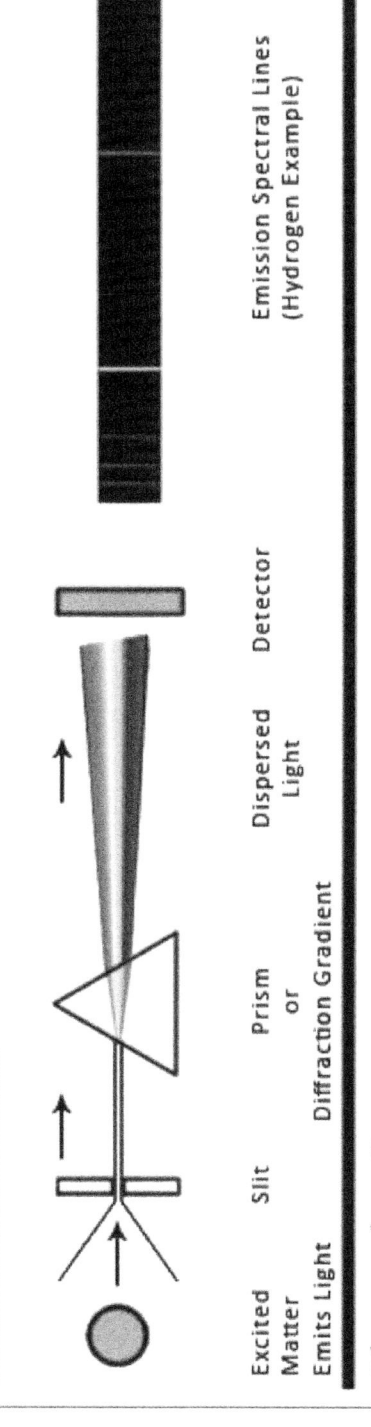

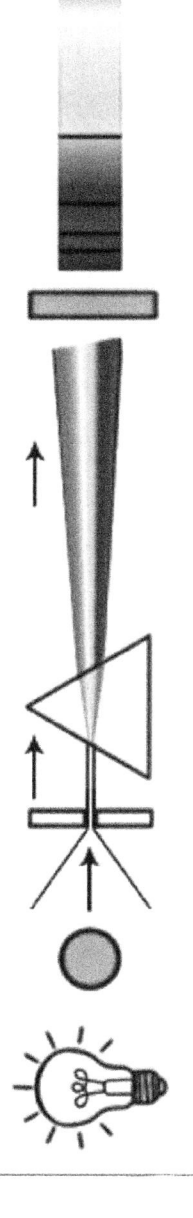

The Symbolism & Significance of Spectroscopy to Science & Survival

Observation occurs when "light" or electromagnetic radiation interacts with matter. Since **observation** is essential to **Science**, and **Spectroscopy** is the study of the interactions between electromagnetic radiation and matter (i.e. the "study of observation") then it makes sense that **Spectroscopy** is one of the fundamental tools used to investigate, inspect, and study matter in the Natural Sciences of Physics, Chemistry, Biology, Geology, and Astronomy. Spectroscopy is a very broad field spanning everything from the subatomic to the astronomic and everything in between. This book serves as an introduction to **Spectroscopy**, so delving in depth into the various sub-disciplines of **Spectroscopy** is beyond the **scope** of this book, however if you are interested, fascinated, or motivated to learn more, you are encouraged to pursue additional research. In particular, it is through the practical application of **Absorption Spectroscopy** which will facilitate the ability to shine electromagnetic radiation on any object or gas you encounter in your day-to-day life to determine the atomic and molecular composition of the object.

Electromagnetic Radiation is a **Fundamental Force in Nature**. "**Visible Light**" is actually a very small portion of the entire Electromagnetic Radiation Spectrum: there are also **Radio Waves, Microwave Rays, Infrared Rays, Ultraviolet Rays, X-rays**, and **Gamma Rays**. "**White Light**" emits all of the color frequencies of the visible electromagnetic radiation spectrum. "**Black Light**", does not emit any color frequencies in the visible electromagnetic radiation spectrum. Therefore electromagnetic radiation is present in both "Light" and "Darkness." The electromagnetic radiation present in "Darkness" is in the form of Microwaves, Radio waves, Infrared Rays, Ultraviolet Rays, X-rays, and/or Gamma Rays. Dark is the absence of visible electromagnetic radiation frequencies, but Dark is NOT the absence of ALL electromagnetic radiation frequencies.

Since the electromagnetic radiation spectrum is "**more than what meets the eye**" in the visible light spectrum, **Spectroscopy** makes use of all of the frequencies of the electromagnetic radiation spectrum to study matter. The **Gamma ray** frequencies of the electromagnetic radiation spectrum are used to study **radioactive atomic nuclei**. The **X-Ray** frequencies of the electromagnetic radiation spectrum are used to study the **atomic electron structures** of matter. The **Ultra-Violet** and **Visible light** frequencies of the electromagnetic radiation spectrum are used to study the **electron transitions** within the atomic and molecular structure of matter. The **Radio** wave frequencies of the electromagnetic radiation spectrum are used in **Astronomy** to study **celestial objects**. The **Microwave** frequencies of the electromagnetic radiation spectrum are used to study **molecular rotations** within matter, and the **Infrared** frequencies of the electromagnetic radiation spectrum are used to study the **molecular vibrations** within matter.

One very important thing to know about all matter is that at the atomic and molecular level, all matter is moving. At the atomic level, there are electrons moving around the nucleus of the atom, and at the molecular level there are **molecular vibrations** between the atomic bonds within the molecule. The atomic bonds within a molecule can vibrate by stretching, wagging, bending, rocking, twisting, or scissoring. The Natural periodic movement, or

Resonant Frequency, that occurs within matter means that matter has its own **vibrational frequency** which serves as a unique fingerprint or signature to identify the type of matter. At the molecular level, the natural **vibrational frequency** of matter occurs within the frequency range of **Microwave** and **Infrared** electromagnetic radiation (0.0003 THz to 400 THz). Molecular Vibrational frequency can be calculated using a formula called **Hooke's Law**. This means that when electromagnetic radiation is **absorbed** by matter, it is due to the **electromagnetic radiation frequency** matching the same **vibrational frequency** of the matter. This is how **Infrared Spectroscopy**, or **IR Spectroscopy**, is able to use electromagnetic radiation to identify specific substances. And since **Infrared Spectroscopy** uses the **Absorption Spectrum** of matter, then the matter is able to be identified without it being potentially destroyed.

In addition to **Infrared Spectroscopy**, another method used to identify molecules based on their vibrations is called **Raman Spectroscopy**. The profound implications and applications of this is that all matter can be identified by its vibrational frequency signature using **Spectroscopy**. In the year 2020, the Human population is being ravaged by the **COVID-19 global pandemic**. If the molecular vibration frequency of **COVID-19** can be identified, then people could use **Spectroscopy** to scan surfaces and air for the presence of **COVID-19** in order to avoid getting sick. In the year 2020, there have also been conspiracy theories surfacing that **5G** (5th Generation) cellular technology is responsible for causing the **COVID-19 global pandemic**. **5G** cellular technology will use electromagnetic radiation frequencies in the range of 600 MHz to 6 GHz, and 24 GHz to 86 GHz.

S P E C - R A - S C O P E

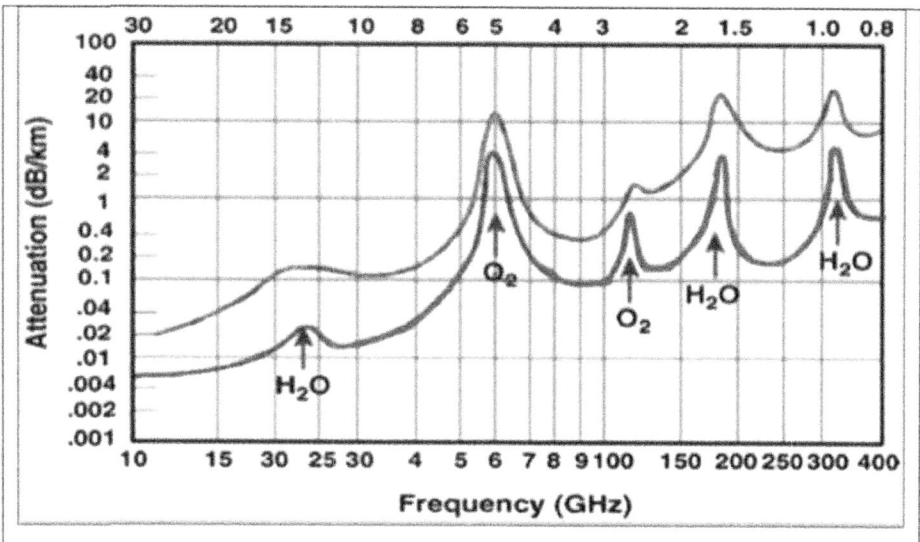

"Average Atmospheric Absorption of Millimeter Waves" from the Federal Communications Commission Office of Engineering and Technology Bulletin Number 70 July, 1997

Promoters the **5G COVID-19 conspiracy** point to a graph from a 1997 bulletin from the **Federal Communications (FCC) Office of Engineering and Technology** that shows that 98% of the **60 GHz** electromagnetic radiation frequency is absorbed by **Oxygen** (the O_2 molecule), and incorrectly interpret this graph as saying that *"98% of the oxygen is absorbed by the 60 GHz frequency and since one of the symptoms of COVID-19 is a shortness of breath, then this is how 5G is causing the symptoms of COVID-19."* Needless to say, the conclusion of the conspiracy theory is based on an incorrect interpretation of the graph. However, comprehending the science of **Spectroscopy** and **molecular vibrations** explains why the **60 GHz** frequency is absorbed by **Oxygen** (the O_2 molecule).

The Symbolism & Significance of Spectroscopy to Science & Survival

The natural molecular vibration of two oxygen molecules atomically bonded (O_2) is a "stretching" type of vibration of around **60 GHz**. Therefore, when a **60 GHz** electromagnetic radiation frequency comes in contact with **Oxygen**, the **60 GHz** electromagnetic radiation frequency is absorbed by the oxygen because the frequencies match; this is how Spectroscopy is able to identify matter by its vibrational frequency. Note that on the FCC graph, there are also frequencies that are absorbed by water (H_2O). The spectroscopic fact that certain molecules absorb and reflect certain frequencies of electromagnetic radiation is how paint is created by adding certain chemical compounds that either absorb or reflect certain frequencies of light to produce the desired color.

Some molecules used to create desired Paint Color		
Molecule	Formula	Paint Color
Carbon	C	Black
Titanium dioxide	TiO_2	White
Cobalt (II) phosphate	$Co_3(PO_4)_2$	Violet
Cobalt (II) aluminate	$CoAl_2O_4$	Indigo
Cobalt (II) stannate	Co_2SnO_4	Blue
Chromium (III) oxide	Cr_2O_3	Green
Zinc chromate	$ZnCrO_4$	Yellow
Cadmium sulfoselenide	Cd_2SSe	Orange
Iron (III) oxide	Fe_2O_3	Red

Mass Spectrometry is another technique used to inspect the interior architecture of molecules. **Mass Spectrometry** and **IR Spectroscopy** are the metaphorical **"two eyes"** of

Spectroscopy utilized frequently in **Organic Chemistry**. In addition to **Atomic Spectroscopy** mentioned earlier, another important spectroscopic technique utilized in Organic Chemistry is called **Nuclear Magnetic Resonance (NMR) Spectroscopy** which provides scientists with information about the electrons around a nucleus. **Atomic Spectroscopy** is also used in **Astronomy** to determine the atmospheric composition of exoplanets. When an exoplanet passes in front of the star that it orbits, some of the star's light passes through the atmosphere of the exoplanet, and Astronomers use Spectroscopy to analyze the star light after it passes through the atmosphere of an exoplanet to determine the composition of the atmosphere of the **exoplanet**.

It is through the use of Spectroscopy that scientists are able to determine which **exoplanets** are potentially habitable and "**Earth-like**." Spectroscopic analysis of the atmosphere of exoplanets can also provide signs of **extraterrestrial life** if any of the elements associated with life are found in the atmosphere of the exoplanet based on spectroscopic analysis. Specifically, **Spectroscopy** used to study celestial objects and the light coming from stars is called **Stellar Spectroscopy** or **Astronomical Spectroscopy**, and most telescopes and microscopes used for research are equipped to perform spectroscopic analysis. **Spectroscopy** not only facilitates the ability to be able to explore our world, but it also provides us with the ability to explore other worlds as well.

5. *Spec*trometry

The etymological origin of the word **"Spectrometry"** comes from the words **"spectro-"** meaning **"spectrum"** (which comes from **"spec"** meaning **"to observe, to view, or look at"**) and the word **"-meter"** meaning **"to measure."** Where Spectroscopy is the "<u>study</u>" of

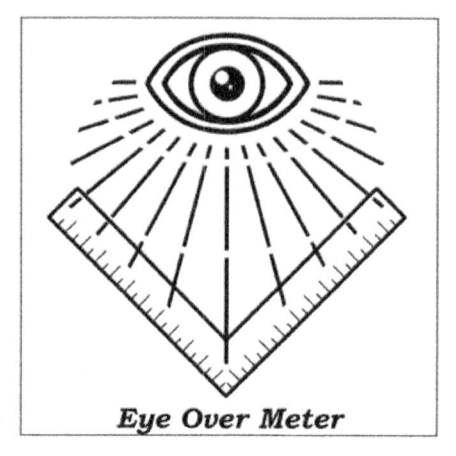

Eye Over Meter

the interactions between electromagnetic radiation and matter, Spectrometry is the "<u>**measurement**</u>" of the interactions between electromagnetic radiation and matter. **Measurement** of the interactions between electromagnetic radiation and matter is needed to **study** the interactions between electromagnetic radiation and matter. Additionally, the science and theory about the interactions between electromagnetic radiation and matter is required to engineer the technology and tools needed to measure the interactions between electromagnetic radiation and matter. Thus **Spectrometry** is needed for **Spectroscopy**, and vice versa. Understanding the difference between **Spectrometry** and **Spectroscopy** is like understanding the difference between **Engineering** and **Science**. Whereas **Engineering** is the practical application of **Science**, **Spectrometry** is the practical application of **Spectroscopy**. Whereas the product of Spectroscopy is theory, the product of Spectrometry is quantitative results.

The primary technologies, tools, or instruments used for **Spectroscopy** to generate and study electromagnetic radiation spectra are the **Spectroscope, Spectrometer**, and **Spectrograph**. All of these tools operate in a similar fashion, but produce different outputs. A **Spectroscope** takes light or electromagnetic radiation as an input and "splits" or "spreads out the light" into its constituent frequencies using a gradient or prism through a process called **dispersion** or **diffraction** producing an observable spectrum. **Spectrometers** operate similar to **Spectroscopes** with the added function of being able to analyze and measure how far the light was spread out after passing through the gradient or prism after being dispersed or diffracted. With a **Spectrometer** you are able to move the scope around the spectrum to view each frequency of light and also measure the **angle of diffraction** to determine the wavelength. Data generated from a Spectrometer can be graphed as **light intensity** on the y-axis versus **light frequency** (or **wavelength**) on the x-axis, and these graphs are referred to as **spectrum** (singular) or **spectra** (plural). Each element or molecule has a unique **spectrum** by which it can be identified by the highest peaks or points on the graph for certain frequencies. Scientific research institutes and organizations maintain databases of the spectra called **Spectral Libraries** for the periodic elements and also various molecules. Using input from a Spectrometer and querying the data in a Spectral Library database will facilitate real-time material identification. A **Spectrograph** operates similar to a **Spectroscope** with the added function of being able to save the output spectrum as a photograph called a **spectrogram**.

The Symbolism & Significance of Spectroscopy to Science & Survival

Spectrometer Schematic

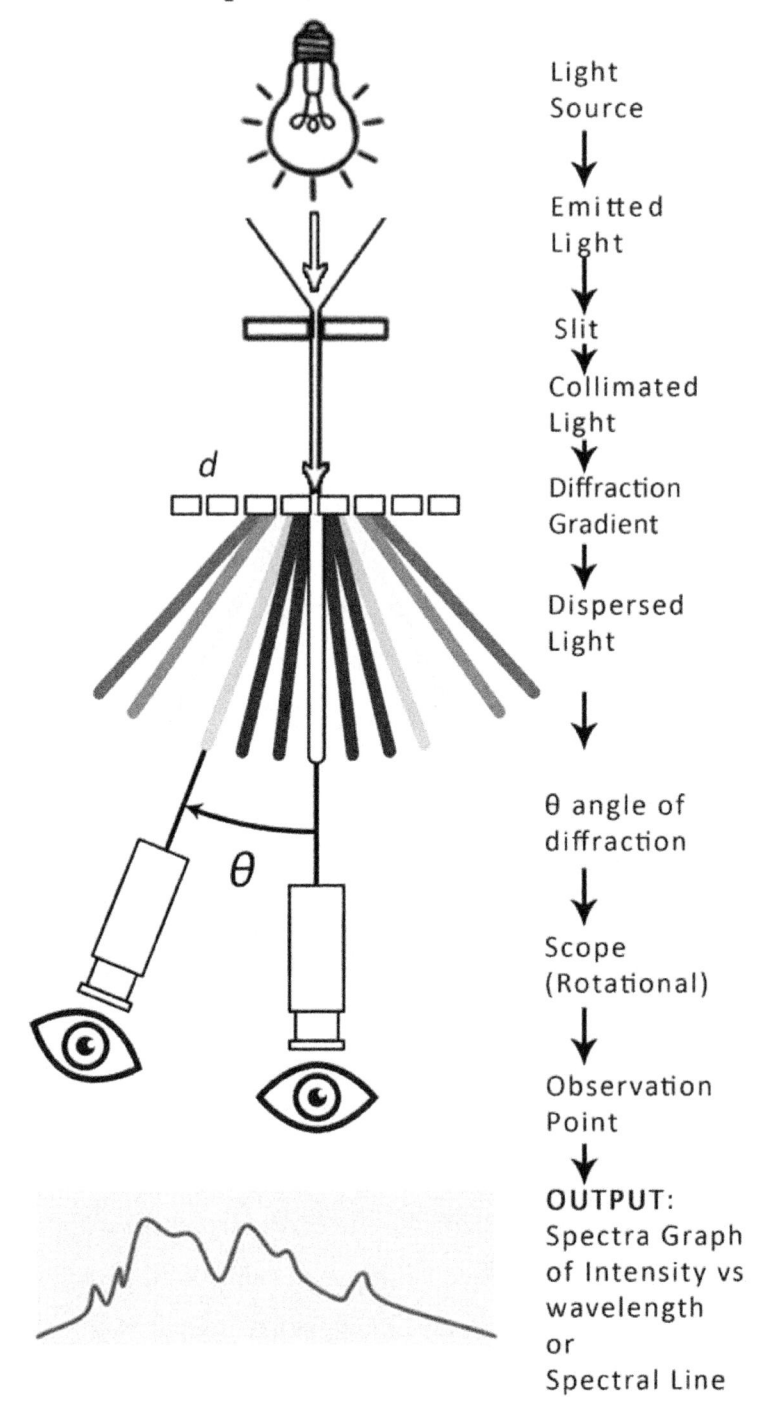

Light Source
↓
Emitted Light
↓
Slit
↓
Collimated Light
↓
Diffraction Gradient
↓
Dispersed Light
↓
θ angle of diffraction
↓
Scope (Rotational)
↓
Observation Point
↓
OUTPUT: Spectra Graph of Intensity vs wavelength or Spectral Line

S P E C - R A - S C O P E

The wavelength of light can be calculated based on the angle measured by the **Spectrometer** using the **diffraction formula** $\lambda = (d*\sin\theta) \div n$, where:

    λ = wavelength

    d = distance between the slits of the diffraction grating

    θ = the angle of diffraction

    n = order number (number of repeating diffraction patterns)

The **frequency** and **wavelength** of light have an inversely proportional relationship. **Frequency (f)** can be calculated if **wavelength (λ)** is known from the equation $f = c/\lambda$ and **wavelength (λ)** can be calculated if **frequency (f)** is known from the equation $\lambda = c/f$ where c is the **speed of light 299,792,458 m/s**. **Light**, or **electromagnetic radiation**, has both **electric** and **magnetic** components as is indicated by the nomenclature. In addition to the **speed of light**, **frequency**, and **wavelength**, another important property of **electromagnetic radiation** is **Intensity**. Electrical **Power** is calculated by the formula $P = I*V$, where **P** represents Electrical **Power**, **I** represents Electrical **Current**, and **V** represents Electrical **Voltage**. The **Intensity of Light** is defined as "**Power per unit area**" or "**Power per square meter**" and thus the **Power of Light** can be expressed as $P = I*A$ where **P** represents Light **Power**, **I** represents Light **Intensity**, and **A** represents **Area unit in square meters**. This expression of the "**Power of Light**" is consistent with the Electrical, Thermal, Hydraulic, and Mechanical interdisciplinary analogies which have been developed over time, and have been routinely included as part of curriculums taught in Science and Engineering programs at Universities across the world. The relationship between Power, Intensity and unit Area in **Spectroscopy**, to the relationship between Power, Current, and Voltage known as **Ohm's Law** in **Electrical Engineering**, to the

The Symbolism & Significance of Spectroscopy to Science & Survival

relationship between Power, Temperature, and Heat Flow known as **Fourier's Law** in **Thermodynamics**, to the relationship between Power, Pressure, and Fluid Flow known as **Poiseuille's Law** in **Fluid Dynamics**, to the relationship between Power, Force, and Velocity known as **Dashpot** in **Mechanical Engineering**, all have a relationship to the **Djed**, **Ankh**, and **Waas** symbols representing **stability**, **life**, and **power**, held by the **Ancient African deity of Engineering** named **PTAH**. Therefore, the symbols of the **Djed**, **Ankh**, and **Waas** can be utilized as a **MASTER KEY** by Science and Engineering teachers, instructors, and professors to students of Egyptian, Kemetic, or African-centered studies, as part of a Transformative learning and Constructivist pedagogy.

"The Master Key" - Interdisciplinary analogies

MASTER KEY	ELECTRICAL	THERMAL	HYDRAULIC	MECHANICAL
WAAS (Power) — POWER (P)	$P = I \times V$ $P = R \times I^2$ $P = V^2 \div R$	$P = q \times T$ $P = R \times q^2$ $P = T^2 \div R$	$P = G \times p$ $P = R \times G^2$ $P = p^2 \div R$	$P = v \times F$ $P = R \times v^2$ $P = F^2 \div R$
ANKH (Life) — CURRENT (I) / HEAT FLOW (q) / FLUID FLOW (G) / VELOCITY (v)	$I = V \div R$ $I = P \div V$ $I = \sqrt{(P \div R)}$	$q = T \div R$ $q = P \div T$ $q = \sqrt{(P \div R)}$	$G = p \div R$ $G = P \div p$ $G = \sqrt{(P \div R)}$	$v = F \div R$ $v = P \div F$ $v = \sqrt{(P \div R)}$
DJED (Stability) — VOLTAGE (V) / TEMPERATURE (T) / PRESSURE (p) / FORCE (F)	$V = I \times R$ $V = P \div I$ $V = \sqrt{(P \times R)}$	$T = q \times R$ $T = P \div q$ $T = \sqrt{(P \times R)}$	$p = G \times R$ $p = P \div G$ $p = \sqrt{(P \times R)}$	$F = v \times R$ $F = P \div v$ $F = \sqrt{(P \times R)}$
PTAH — RESISTANCE (R) / THERMAL RESISTANCE (R) / FLOW RESTRICTION (R) / FRICTION (R)	$R = P \div I^2$ $R = V^2 \div P$ $R = V \div I$	$R = P \div q^2$ $R = T^2 \div P$ $R = T \div q$	$R = P \div G^2$ $R = p^2 \div P$ $R = p \div G$	$R = P \div v^2$ $R = F^2 \div P$ $R = F \div v$

6. *Spec*ifications

The etymological origin of the word "Specifications" comes from the word "specific." The symbol of the "**Bullseye**" or "target" is used as an icon to represent or indicate something "specific." The word "specific" comes from the word "species," which comes from the Proto-Indo-European

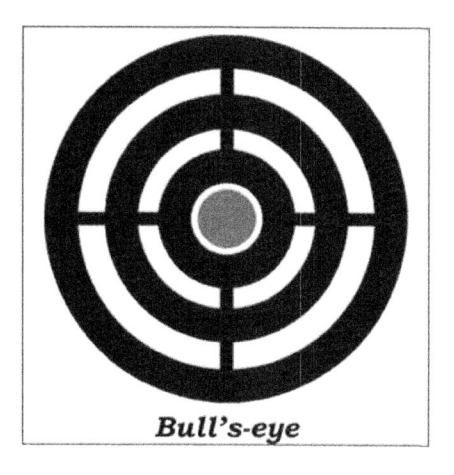

Bull's-eye

root "spek-" meaning "to observe." In the field of Technology, "**Technical Specifications**" or "**Tech Specs**," refers to the documentation of the specific methods and procedures of a project or product. In this chapter two "**experiments to experience evidence**" of the principles of **Spectroscopy** will be presented. The first experiment will involve constructing a simple **Spectroscope** and the second experiment will involve constructing a simple **Spectrograph** which will produce **spectrograms**. Since special tools are needed to detect electromagnetic radiation outside of the visible range, and the **Absorption Spectrum** within the visible range can be determined by the color of objects that you see every day, both the **Spectroscope** and **Spectrograph** described in the following experiments can be used to investigate the **Emission Spectrum** of **Electromagnetic Radiation** in the **visible** range.

The Symbolism & Significance of Spectroscopy to Science & Survival

The following are light sources that can be viewed using the **Spectroscope** and **Spectrograph** projects in this chapter:

1. **Noon and Rising/Setting Sun**: Spectrum = Continuous spectrum with dark lines where certain frequencies are absorbed by the Earth's atmosphere. The Rising/Setting Sun will have more dark lines than the spectrum produced by the Noon day Sun because when the Sun is lower, the light passes through more of the Earth's atmosphere, and so more frequencies are absorbed. *(Note: do not view the sunlight directly, but rather view the light reflected off of a white piece of paper or white wall)*

2. **Moon Light:** Spectrum = Continuous. Moonlight is reflected Sunlight so the spectrum should look similar

3. **Incandescent Light**: Spectrum = Continuous spectrum produced by the thermal blackbody radiation emitted by exciting a filament made of the element Tungsten.

4. **Fluorescent Light**: Spectrum = Discrete Spectral lines due to the excitation of the element Mercury.

5. **Candle Flame**: Spectrum = Continuous spectrum produced by thermal blackbody radiation emitted by burning the candle wick. The type of wick will determine the spectrum.

6. **Yellow Street Lights**: Spectrum = Discrete Spectral lines due to excitation of the element Sodium

7. **Neon Light**: Spectrum = Discrete Spectral lines due to excitation of the element Neon

8. **Colored LED Light**: Spectrum = Continuous spectrum of only one color

9. **Computer Monitor**: Spectrum = Discrete Spectral lines of Red, Green, and Blue

S P E C - R A - S C O P E

EXPERIMENT 1: Create Your Own Spectroscope

WARNING: This experiment involves using scissors and a razor to cut paper. Proceed with caution! Minors should get the permission and supervision of a parent or guardian prior to performing this experiment. The Author is not liable for accidents that occur while performing this experiment.

Materials Needed:
- 2 pieces of 8.5in x 11in Black cardstock paper
- Scissors, razor, or X-acto knife
- Ruler
- Drawing Compass
- Glue
- 1 CD-R or DVD-R

Instructions:

1. Step 1: Prepare one piece of 8.5in x 11in Black cardstock paper by drawing two vertical lines the full length of the paper 2 inches from the edge of the paper. Draw another vertical line in the center of the paper half the length of the paper to the bottom of the paper.

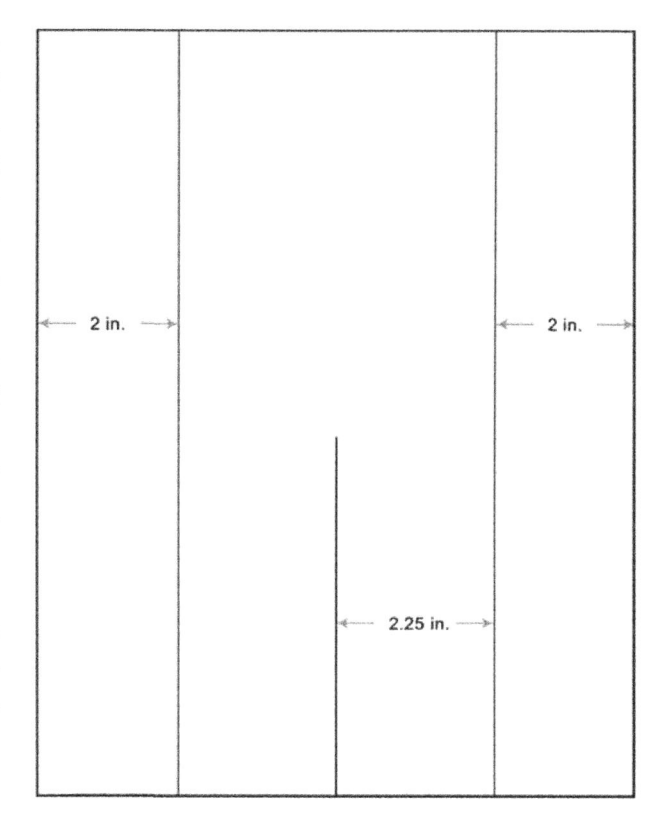

93

2. <u>Step 2</u>: Continue preparing the 8.5in x 11in Black cardstock paper by drawing two horizontal lines the full width of the paper. One line should be 6.5in from the top of the paper and the other line should be 4in from the bottom of the paper. Within these two horizontal lines, centered within the 2 inch margins you drew in Step 1, draw two 0.5in x 0.5in squares. Draw a dot to indicate the following 3 points: 1) the intersection of the left vertical line and the lower horizontal line, 2) the intersection of the center vertical line and the top horizontal line, and 3) the intersection of the right vertical line and the lower horizontal line.

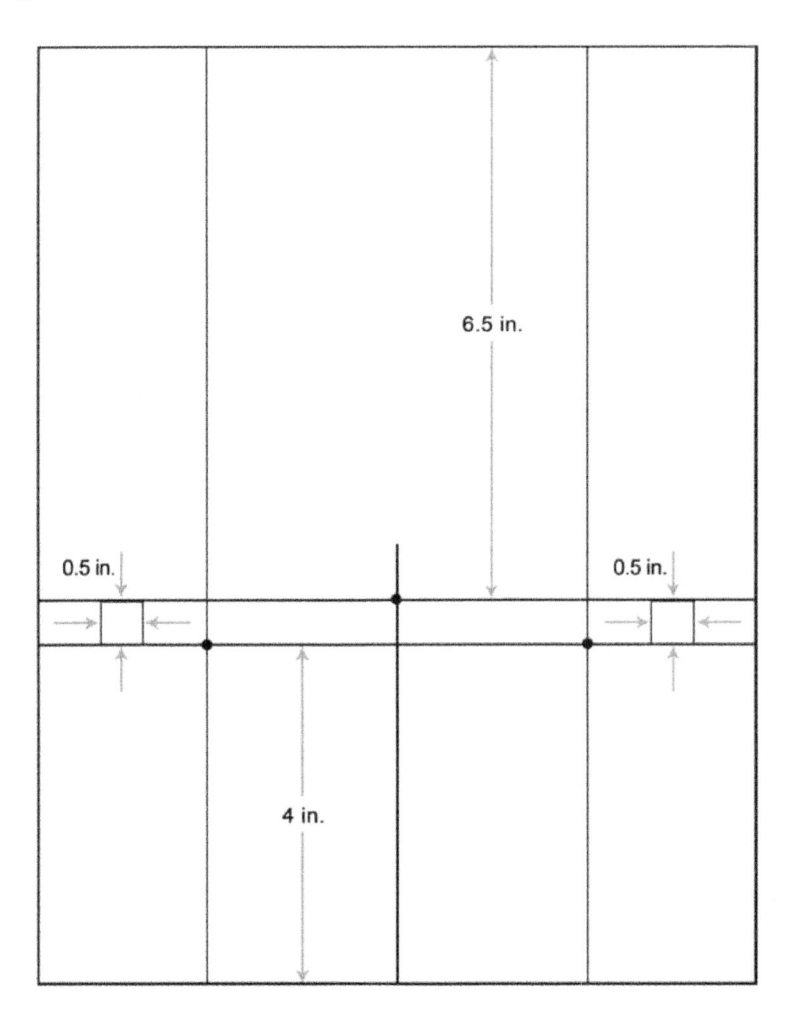

SPEC - RA - SCOPE

3. <u>Step 3</u>: With your drawing compass centered on the bottom of the paper, draw an arc connecting the three points identified in Step 2: 1) the intersection of the left vertical line and the lower horizontal line, 2) the intersection of the center vertical line and the top horizontal line, and 3) the intersection of the right vertical line and the lower horizontal line.

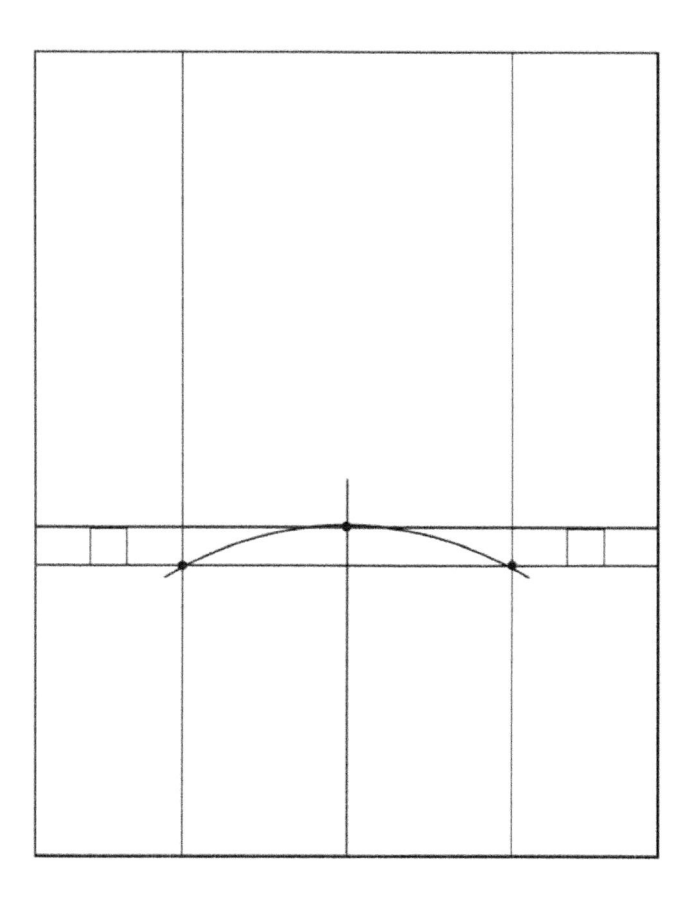

4. <u>Step 4</u>: Use your Scissors, razor, or X-acto knife to cut out the two 0.5in x 0.5in squares (this will be the viewing hole). Also cut the paper along the arc you drew in Step 3 (this is where the CD/DVD will be inserted)

The Symbolism & Significance of Spectroscopy to Science & Survival

5. Step 5: On the other piece of 8.5in x 11in Black cardstock paper, draw two circles of diameter 2.1 inches. In the center of one of the circles, draw a rectangle with dimensions 4cm long by 2mm wide. Use your Scissors, razor, or X-acto knife to cut out the two circles, and cut the 4cm by 2mm rectangle out of one of the circles.

6. Step 6: Put glue within the borders of one of the 2 inch margins you drew in Step 1. Roll the prepared 8.5in x 11in Black cardstock paper such that the two 0.5in x 0.5in squares overlap to form one 0.5in x 0.5in viewing hole into the tube. When glue on the tube dries, glue the two 2.1in circles created in step 5 to the tube. The circle without the slit should cover the bottom of the tube closest to the 0.5in x 0.5in viewing hole. The circle with the slit should cover the top of the tube farthest from the 0.5in x 0.5in viewing hole. The circle should be oriented such that the slit in the circle is horizontal to the 0.5in x 0.5in viewing hole.

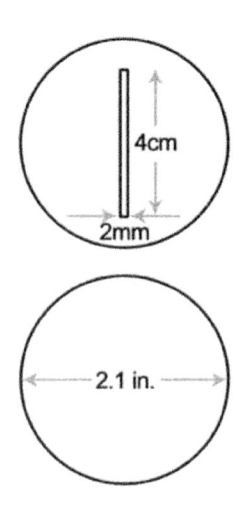

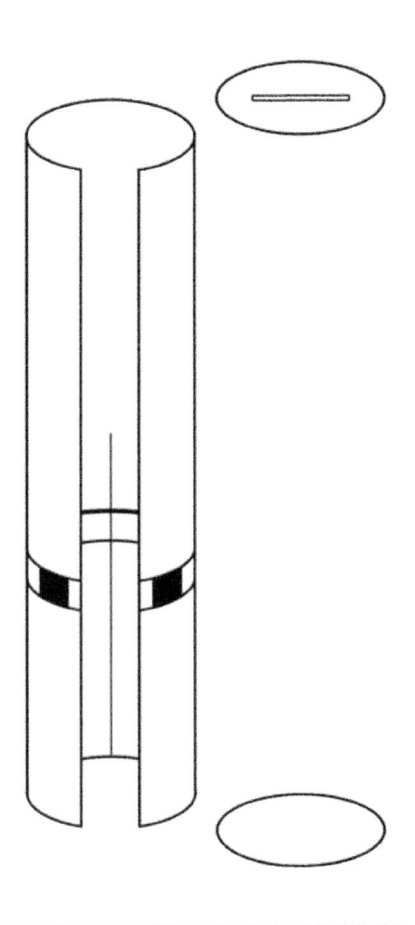

SPEC - RA - SCOPE

7. Step 7: Place a CD or DVD into the arc slit that you cut in step 4. The CD/DVD should be oriented such that the dull side of the CD/DVD that is printed on with writing is facing away from the slit in the top circle of the tube, and the shiny side of the CD/DVD with the "rainbow colors" should be facing up towards the slit in the top circle of the tube. The Spectroscope is now constructed. Allow light to shine through the slit in the top circle of the tube, and view the spectrum of the light after it contacts the diffraction grating of the CD/DVD through the 0.5in x 0.5in viewing hole.

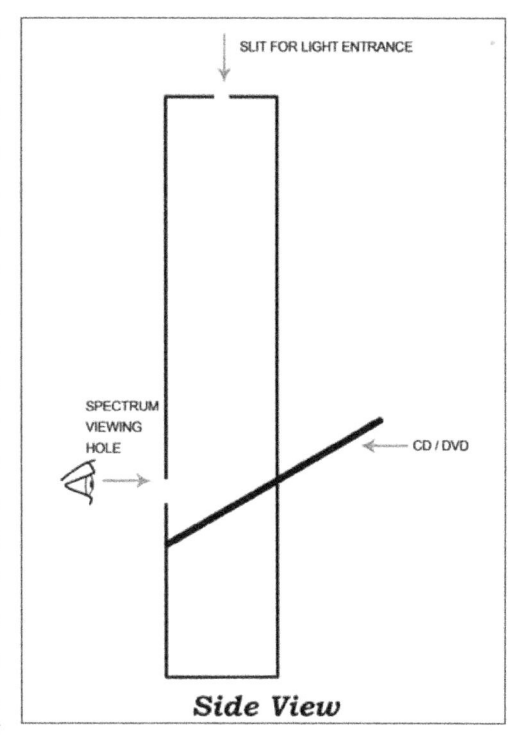

Side View

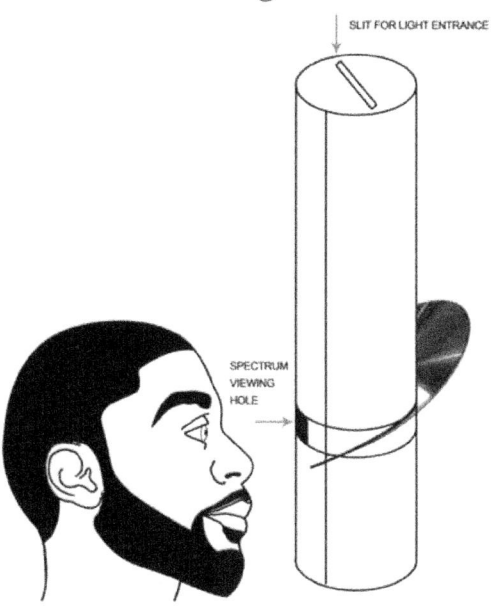

97

The Symbolism & Significance of Spectroscopy to Science & Survival

EXPERIMENT 2: Create Your Own Spectrograph

WARNING: This experiment involves using scissors and a razor to cut paper and plastic. Proceed with caution! Minors should get the permission and supervision of a parent or guardian prior to performing this experiment. The Author is not liable for accidents that occur while performing this experiment.

Attribution: This project was developed by **Public Lab contributors (PublicLaboratory.org)** and is reproduced, shared, and distributed here under the CERN Open Hardware License 1.1.

Materials Needed:
- 1 piece of 8.5in x 11in Black cardstock paper
- Scissors, razor, or X-acto knife
- Tape
- Glue
- 1 CD-R or DVD-R
- Cell phone or laptop with a camera

Instructions:
1. Step 1: Cut out the pattern on **page 101** and trace it onto Black cardstock paper. Cut along the outer edge. Fold up or down as indicated by the dotted and dashed lines.

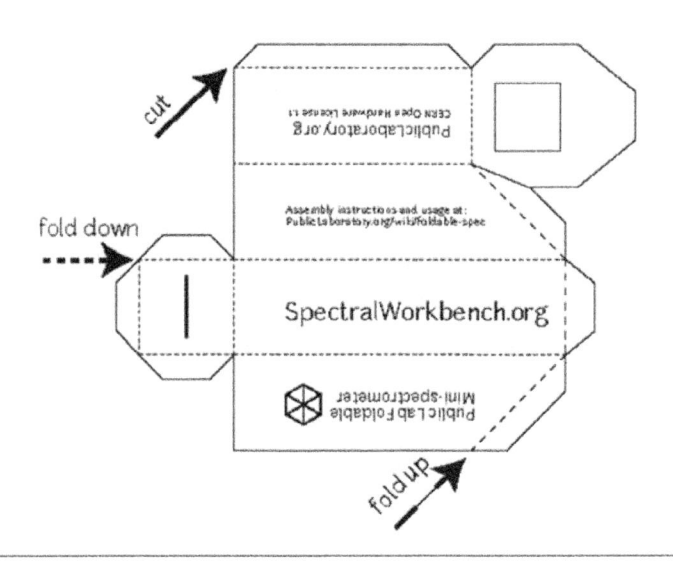

SPEC-RA-SCOPE

2. <u>Step 2</u>: Except for the diffraction grating door, glue or tape all flaps down onto the outside.

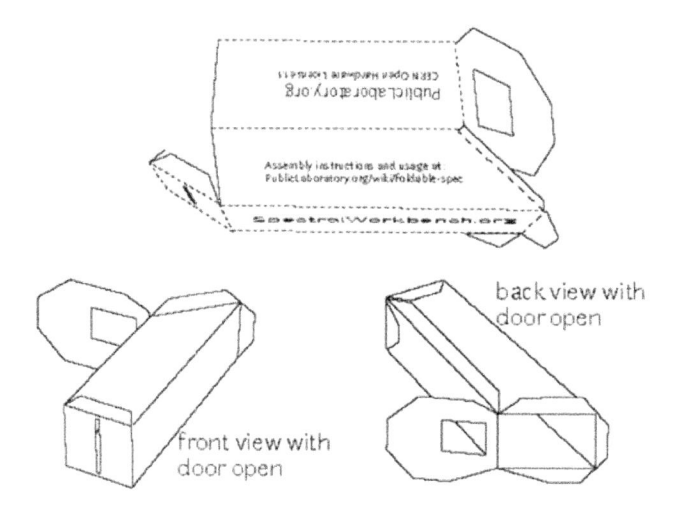

3. <u>Step 3</u>: Make a diffraction grating from a CD-R or DVD-R by splitting it into quarters, then peel off the reflective layer and trim a small clean square out of the transparent layer. Use a piece of the CD-R or DVD-R without fingerprints or scratches.

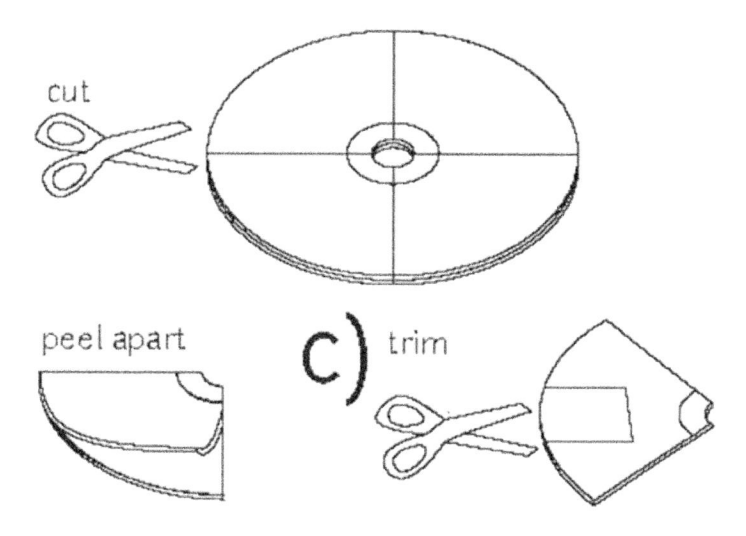

The Symbolism & Significance of Spectroscopy to Science & Survival

4. Step 4: The CD-R or DVD-R diffraction grating piece must be placed so that its grating is vertical, making a horizontal spectral rainbow. Tape your CD-R or DVD-R piece to the inside of the diffraction grating door of the Spectrograph, then tape or glue the door closed.

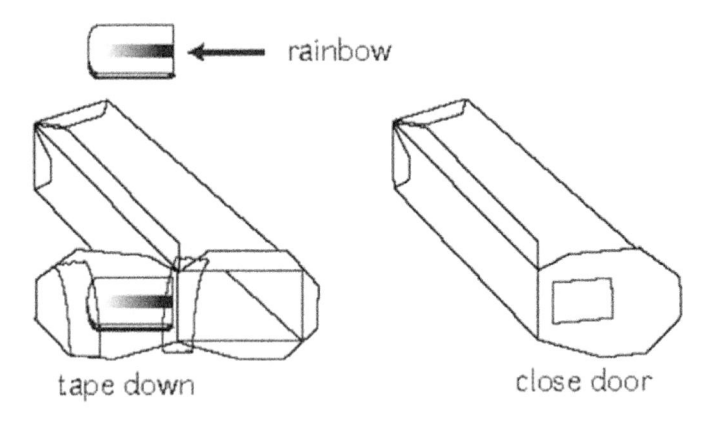

rainbow

tape down   close door

5. Step 5: The Spectrograph can be mounted on a camera phone, a laptop with a camera, or a webcam. Line up carefully so that the rainbow is in the middle of the image, and tape down firmly so that the Spectrograph stays rigid. The Spectrograph can now be used to take Spectrogram pictures of spectrum from various light sources. More assistance and details about this project can be found on the following websites:
https://publiclab.org/wiki/foldable-spec
https://spectralworkbench.org/

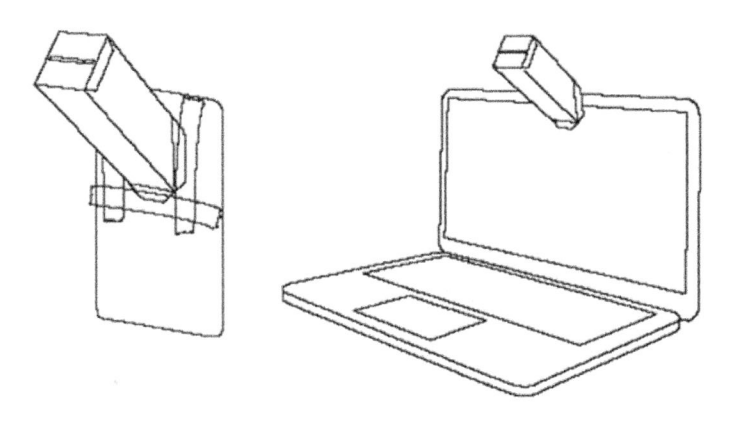

SPEC-RA-SCOPE

1 in x 1 in

Public Lab Foldable
Mini-spectrometer

SpectralWorkbench.org

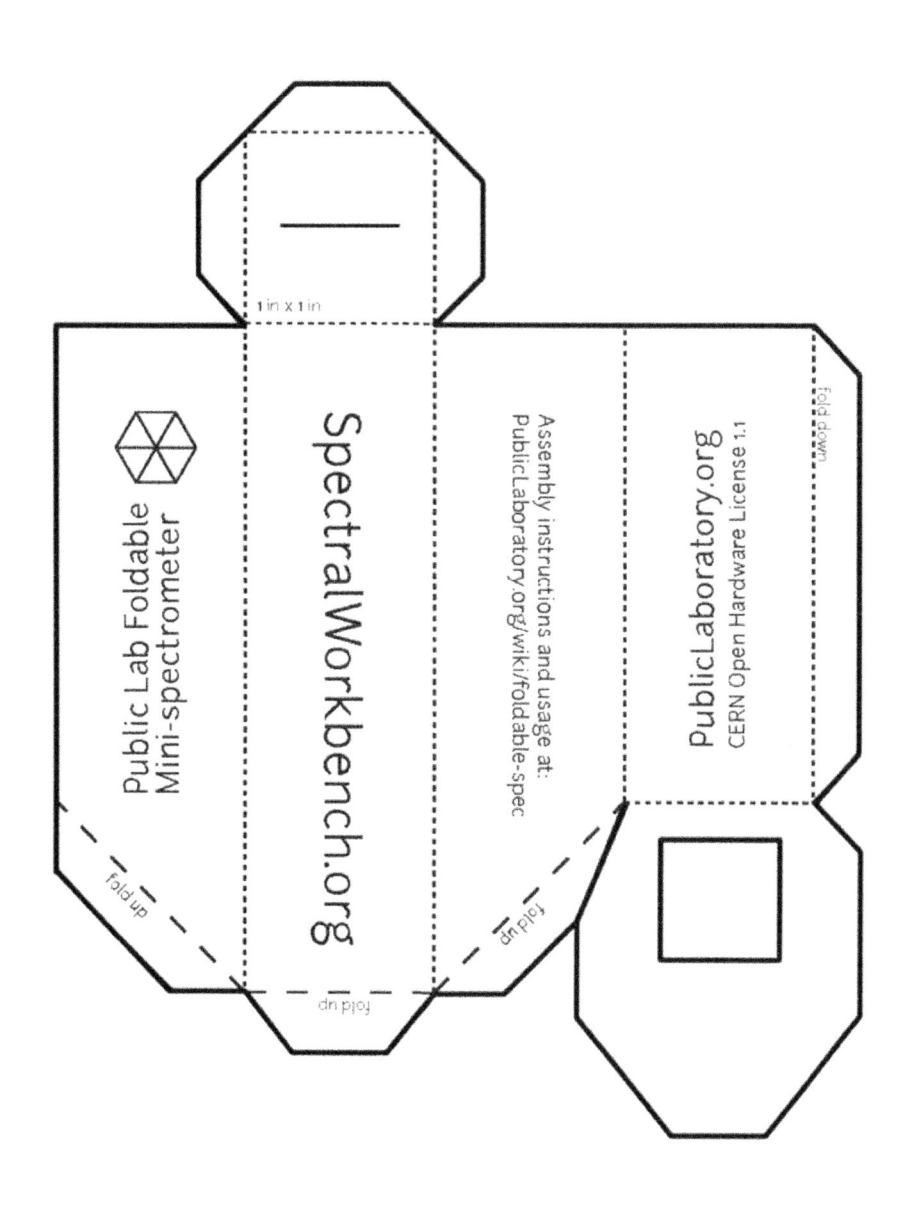

Assembly instructions and usage at:
PublicLaboratory.org/wiki/foldable-spec

PublicLaboratory.org
CERN Open Hardware License 1.1

fold up

fold up

Fold down

7. Re*spec*t

The etymological origin of the word "**Respect**" comes from the prefix "**Re-**" meaning "back," and the root word "**spec**," which as previously mentioned, means "**to look or to observe**." So, although the definition of the word "**Respect**" means "*to hold in high regard or to hold in high esteem*," the etymological origin of the word "**Respect**" implies a

Sankofa –
West African symbol of Respect for the past

meaning of "**to look back**." Therefore, combining the definition and etymology of the word "Respect" gives the connotation of "*looking back at something that is held with high regard*." While the English words "**Review**" and "**Retrospective**" both convey meanings of "looking back," it is the dual functionality of the word "Respect" referring to both "looking back" and "honoring" simultaneously, which makes it the appropriate word to describe the conclusion of this book on **Spectroscopy Science for Survival**. In the West African Adinkra hieroglyphic (sacred symbols) system, the **Sankofa** symbol embodies the same dual meaning and function as the English word "Respect." The **Sankofa** symbol is depicted as a bird with its head turned backwards. In the Twi language spoken in **Ghana** West Africa, the word "**Sankofa**" means "go back and get it". The interpretation and application of **Sankofa** is that of "**Respect**" by looking back on the

103

knowledge, accomplishments, and achievements made in the past, and honoring the past by making the past relevant in the modern day. The Sankofa symbol represents a practical approach to studying History by taking from the past what is good, and bringing it into the present, in order to make positive progress through the practical application and utilization of knowledge. **"Sankofa"** has been the pedagogical method utilized throughout all of the **African Creation Energy** books by taking African cultural symbols and concepts from the past and making them relevant and practical in the here-and-now in the fields of **Science**, **Technology**, **Engineering**, and **Mathematics**.

Popular Astrophysicist and Science Communicator **Neil DeGrasse Tyson** has often spoken about the fact that the right to name scientific concepts should belong to the person, group of people, or culture who discovers the concept. He points to the fact that the reason why there are so many stars with Arabic names is because there was a point in time in history where members of the Arab culture were very active in identifying and naming the stars. He also points to the fact that the reason why there are elements on the periodic table with names like **Californium** and **Berkelium** is because these elements were discovered in the city of Berkeley, California in the U.S.A. However, when we observe the historical contributions and discoveries by different cultures and groups of people in the fields of Science, Technology, Engineering, and Mathematics, we do not always see a correspondence of the **"Naming Rights"** of scientific concepts belonging to the culture or group of people who discovered the concept. For example, in the field of Astronomy, the names of Greco-Roman deities are used as the names of planets in our Solar System: **Mercury, Venus, Mars, Jupiter, Saturn, Uranus**, and **Neptune**. However, the historical evidence points to the fact that the **Ancient Mesopotamians** of Babylon were the first to utilize observational Astronomy and provide names of the first five planets in our Solar System which were **Nebo** (Mercury), **Ishtar** (Venus), **Nergal** (Mars), **Marduk** (Jupiter), and **Ninurta** (Saturn). So, since the Ancient Babylonians were the first to discover and name these concepts, why do we not **Respect** their discovery by utilizing the names they provided?

When we study the history of the field of Mathematics and we learn that the **Lebombo Bone** found in the **Kingdom of Eswatini (Swaziland)** in **Africa** is the oldest mathematical object in the world with evidence of applied **Arithmetic**, then we should **Respect** their discovery and contribution and name the field of **Arithmetic** after an African term in the language of **Eswatini (Swaziland)**! Even the numbers that we use in Mathematics which are called "**Arabic Numerals**" are not actually Arabic at all but are derived from **Kabylia** in **Algeria** in **North Africa**. So, we should **Respect** the contribution of that culture and group of people by calling the numbers "**African Numerals**" or "**Algerian Numerals**". If the earliest example of the **Scientific Method** can be found on a Medical Papyrus from **Ancient Egypt** (called the "*Edwin Smith Papyrus*"), then we should **Respect** the contribution of the Ancient Egyptians by using the term from the Medu Neter language that the Ancient Egyptians used for the **Scientific Method** which was **Tep Hesp** (meaning "*correct method*"). When we see applied evidence of the **Pythagorean Theorem** in the size and construction of the **3 Pyramids of Giza** as a "**Pythagorean Triple**" thousands of years before Pythagoras was said to have lived, then we should **Respect** the Ancient Egyptian's knowledge and application of this concept and name the formula $a^2 + b^2 = c^2$ after a term from **Ancient Egypt.**

To date, all cultures and races of people have contributed to the body of knowledge that is Science. Thus, scientific terminology should reflect the spectrum of all cultures and races of people. This proposition is not to disrespect or take away from the contributions that White people and the nations of Europe have originally and uniquely made to science, but rather to promote equality and pay respect to all of the original and unique contributions that all cultures and races of people have made to Science over time. Currently there is bias built into the presentation and education of Scientific information. And that bias and exclusion creates and perpetuates a sense of superiority amongst certain groups of people, and a sense of inferiority amongst other groups of people, and why many Black African people feel separate and apart from Science because they do not see where Black African people contributed anything to the S.T.E.M. fields. To not give the different cultures and groups of people the **Right** and **Respect** to name the Scientific concepts which they discovered, and to omit their contributions in favor of a narrative that predominately celebrates one culture even when there is evidence that other cultures made certain scientific discoveries first, is not Scientific, but rather **Pseudoscientific**. When certain information or data is omitted or left-out in order to serve a certain narrative, that is the methodology of a **Pseudoscientist**. It is the omission of the historical contributions of Black African people to the S.T.E.M.

fields which leads to Black people occasionally rejecting science and scientific concepts calling it "***The White Man's Science***".

If there is such a thing as "***White Man's Science***" and "***White Man's Technology***," then there is also such a thing as "***Black Man's Science***" and "***Black Man's Technology***," or "***African Science***" and "***African Technology***". Personally, I prefer the terms "***African Science***" and "***African Technology***" because it is genderless and encompasses the contributions of all people of African descent. Let us recall that technology is just not tools and machines, but it also refers to techniques, skills, processes, and methods. The technology or method that is the foundation which underpins all science is the "**Scientific Method**" which is an African invention developed by African Engineers. As previously mentioned, the earliest example of the **Scientific Method** can be found on a papyrus from **Ancient Egypt**. So, since the "**Scientific Method**" is an African invention and is the foundation of ALL Science, and all people descend from Africans, then technically ALL Science can be considered "*African Science*," the "**Science of Sciences**".

The juxtaposition of these concepts of "***The White Man's Science***" versus "***The Black Man's Science***" falls under the heading of "***Western Science* versus *Traditional Indigenous Knowledge***" and is studied in a field called **Ethnoscience** which studies how different cultures of people develop with different forms of knowledge and beliefs, and focuses on the historical contributions different cultures of people have made to science. The method and means by which scientific concepts are expressed are essentially technologies themselves.

Historically these technologies used to express scientific concepts differed by region and race; and this is studied in the field of Ethnoscience. Mathematics is a technology used to express logical abstract thought. The regional and racial differences in the technologies used to express logical abstract thought (mathematics) is studied in the field of **Ethnomathematics**. Specifically, **Ethnomathematics** is the study of the relationship between mathematics and culture, and also studies the different methods of mathematics which are practiced among different identifiable cultural groups.

For Pan-Africanist and people who promote the idea that Black African people should not worship White Gods or White deities, there is a great work (***magnum opus***) that must be done in the S.T.E.M. fields because when White people made their contributions to the body of knowledge that is Science, they made a point to use their mythology and their names as the names of Scientific concepts even if they were not the first to discover the concept. You may have stopped worshipping White Jesus, but you can't name the planets of the Solar System, or the Days of the Week without calling on European Gods. You can't refer to natural phenomena like Thunder, which means Thor's Din, without calling on White Gods. You can't name the elements of the periodic table without calling on White Gods: 1. Hydrogen, 2. Helium, 34. Selenium, 41. Niobium, 61. Promethium, 73. Tantalum, 80. Mercury, 90. Thorium, 92. Uranium, 93. Neptunium, and 94. Plutonium. And those are just the ones named after White Gods, there are others named after White people. You can't make reference to Units of Measurement like Volt, Amp, and Ohm, without calling on the names of dead White men. Anytime you use the word Thesis,

Theory, Theorem, Chronology, Archeology, Psyche, Psychology, or Astro- anything, you are calling on White Gods. They put their mythology in Science. But when we use our African mythology or our African symbols in Science, we get criticized by our own people. When they named the planet Mercury, did they say, well there is no historical significance in the mythology of the God Mercury which makes reference to that planet so we really should not name it Mercury. Or, did they identify a loose enough association between the planet and the mythological god Mercury to call it Mercury, and now you cannot make reference to the planet without calling on one of their Gods. Historically, it has been the practice of the nations of Europe to claim "discovery" of things originally discovered by, known to, or owned by the indigenous peoples of the world. We need only to look to the example of **Christopher Columbus** and the so-called "discovery" of the "New World" America. While the lands that have become known as the Americas (North America and South America) may have been a new discovery to Europeans, the lands were already discovered and inhabited by indigenous people. It also has been the practice of the Nations of Europe to force their deities and religions onto the native indigenous people they colonized. Why do we not respect the original native indigenous people who discovered the lands of the Americas and call the land by its indigenous name? In the same way Europeans have usurped, supplanted, and colonized the land, they have "colonized the mind" and **"colonized Science"**.

As conscious, "woke", and active Black people of African descent, we should use our own mythology and our own symbols to identify scientific concepts. This has been one of the main purposes of the **African Creation Energy** series of books. A practical example of using African terminology and symbols in Science was demonstrated by Physicist **Dr. Sylvester James Gates Jr.** where he used the African symbol terminology "Adinkra" from West Africa to describe *"A Graphical Technology for Supersymmetric Representation Theory"*. One argument and criticism that I have received for my effort to relate African symbols to Scientific concepts is that the effort is **"Sufficient but not Necessary."** And in terms of the logic of the argument, I agree it is not necessary because there are plenty of examples of Black African people around the world over time who have learned and mastered scientific concepts without ever having the concept related to an African symbol, African word, African mythology, or African deity. However, here I would like you to consider this rhetorical question: *"If the racial depiction of Jesus does not matter, then why is Jesus depicted as White?"* Jesus could and should just as well be depicted as Black, right? "If the type of symbols and mythology used in the S.T.E.M. fields does not matter, then why do we have to use European symbols, especially in the cases where the concept or idea was first discovered by non-Europeans?" Given the amount of mythology from European cultures that is used as symbolism within the S.T.E.M. fields, I think there would be even more Black African people who would gravitate towards the S.T.E.M. fields if African symbology and African mythology was used to express Scientific concepts. Additionally, given the reality of how religious Black African people are globally, and given the fact that there is a disconnect between Science and Religion to the point where some people are unable to think Scientifically due to their preexisting religious programming, I think that relating Scientific concepts to a God concept is both a

Necessary and Sufficient condition for taking the first steps towards developing a more Scientifically literate population.

In the year 2020 in America and around the world, there has been a calling for **Racial Justice** and a removal of statues and symbols of Racism and "**White Racial Preference**" which contribute to a false narrative of "**White Supremacy**". The popular idiom "**hindsight is 20/20**" refers to the fact that as we move forward in time, the best choice or decision is not always apparent, however, when we look back on the situation and evaluate the past, the best option is obvious and clear like **"20/20" vision**. With hindsight being "20/20" and the year **2020** now being mostly over and in our hindsight, this provides us with the opportunity to take a look at similar racial injustices and "White Racial Preferences" which exist in the S.T.E.M. fields. This may be a good sign that in time, the people of Earth will call for the same **Reckoning** in the S.T.E.M. fields, and to **Respect** the S.T.E.M. discoveries and contributions of all the cultures and groups of people on Earth. When this is applied, then as a student matriculates through their various Science and Math classes in school, then they would learn and hear about concepts which have names from every culture and group of people on Earth depending on who was the first to discover the concept. This would empower every Race and Culture to see themselves in the S.T.E.M. fields and see the contributions their culture has historically made to the S.T.E.M. fields. **Representation Matters** and this is proper **Respect** and **Scientific Reparations**.

Now close your eyes, and picture in your mind's eye a future where Africa is totally liberated, where African people are able to make use of their natural resources for their own benefit and enjoy the wealth and prosperity that comes with it, on par with, or greater than that of the other nations of the earth. Picture a future where Africa is self-sufficient and has complete agency in self-determination, where people of African descent are no longer oppressed and suppressed around the world but rather **respected** and held in high esteem in any country where they visit or reside. This future for Africa and people of African descent is indeed possible, and it starts with the acquisition, mastery, and practical application of **Science**, **Technology**, **Engineering**, and **Mathematics**.

> "We've got some difficult days ahead, but it really doesn't matter with me now, because I've been to the mountain top. Like anybody, I would like to live a long life, longevity has its place, but I'm not concerned about that now. I just want to do God's will. And, he has allowed me to go up to the mountain, and I've **looked** over, and I've **seen** the promise land. I may not get there with you, but I want you to know tonight that we as a people will get to the promise land. So, I'm happy tonight, I'm not worried about anything, I'm not fearing any man. Mine **Eyes** have **seen** the glory of the coming of the lord."
>
> ~ Dr. Martin Luther King Jr.

Keep your <u>EYES</u> on the Prize, Hold On, Hold On!

8. In*spec*tor (About the Author)

African Creation Energy can scientifically be defined as the Work, Effort, Endeavors, and Activities of African people that cause a movement or change. **African Creation Energy** is The Energy, Power, and Force that created African people and that African people in turn use to Create. Since African people are the Original people on the planet Earth, it follows from thermodynamics that the Creation Energy of African people is the closest creation Energy of all the people on the Planet to the **Original Creative Energies** that created the Planets, stars, and the Universe. **African Creation Energy** is **Black Power** in the scientific sense of the word "Power", and this book **Radiates African Creation Energy** to be absorbed by the **Black Body**. **African Creation Energy** has been called by many different names amongst many different groups of African people throughout time. **African Creation Energy** has been called by the names Ashe, Tumi, Dikenga, Nyama, Nzambi, Amma, Sekhem, NoopooH, and Nuwaupu just to name a few. The conduit of "**African Creation Energy**" who has written and authored this book, and other books, goes by

The Symbolism & Significance of Spectroscopy to Science & Survival

the title of **Osiadan Borebore Oboadee** from the Twi language spoken in Ghana West Africa. The Twi word **"Osiadan"** comes from the root words "Si" meaning "Build" and "adan" meaning "Building" with "O-" being a way to denote a "Master". Hence "Osiadan" literally describes a **"Master Builder"**. Also note the phonetic similarities between the Twi words "Si" and "Adan" and the Ancient Egyptian words **"Sia"** (wisdom) and **"Aton"** (high noon sun). The Twi word **"Borebore"** comes from the root words "Bo" meaning "Create" and "Re" meaning "to do repetitiously", thus "Borebore" is used to describe a **"Perpetual Creator"**. The word "BoreBore" or "Bore" in Twi is also related to the Hebrew word **"Bara"** meaning **"to begin"** found in the first verse of the first chapter of the Judeo-Christian Bible, and is also related to the Yoruba word "bere" meaning "to begin". Also note the phonetic similarities between the Twi words "Bo" and "Re" and the Ancient Egyptian words "Ba" (soul) and "Re" (sun). The Twi word **"Oboadee"** comes from the root words "Bo" meaning "Create" and "Abode" meaning "Creation" with "O-" being a way to denote a "Master", hence "Oboadee" literally describes a **"Master Creator"**. Oboadee is also pronounced O-Poatee in different African dialects, and is said to derive from the pronunciation of the name of the Ancient African Creation deity **PTAH**. Osiadan, Borebore, and Oboadee are three principles of **African Creation Energy**.

Osiadan is African by blood and lineage; a descendant of the **Balanta-Bassa** and **Djola-Ajamatu** tribes in present day **Guinea-Bissau** West Africa. Both the Balanta and Djola tribes migrated to West Africa in Ancient times from the area which is present day **Egypt**, **Sudan**, and **Ethiopia**. Osiadan is a descendant of the Ancient **Napatan**, **Merotic**, **Kushite** Pyramid Builders, and is a Scientist, Engineer, Mathematician, Problem Solver, Analyst, Synthesizer, Artist, Craftsman, and Technologist by education, profession, and Nature. Osiadan has obtained Bachelors and Masters Degrees in the areas of Electrical Engineering, Physics, and Mathematics. Born in the African Diaspora, Osiadan made his first trip to the African continent in the year 2008. Between the years of 2009 and 2010, Osiadan Borebore Oboadee set out to develop, engineer, invent, formulate, build, construct, and create several Technologies (Applications of Knowledge) for the well being of African people worldwide and attempted to radiate the energy that motivated and inspired the development of those technologies in a three part introductory educational series which collectively was entitled "The African Liberation Science, Math, and Technology Project" **(The African Liberation S.M.A.T. Project)**. The three books that are part of African Creation Energy's "African Liberation S.M.A.T. project" are:

1. **SCIENCE:** (Knowledge/Information)
   The SCIENCE of Sciences, and The SCIENCE in Sciences
2. **MATHEMATICS:** (Understanding/Comprehension)
   9^{9^9} Supreme Mathematic African Ma'at Magic
3. **TECHNOLOGY:** (Wisdom/Application)
   P.T.A.H. Technology: Engineering Applications of African Science

The primary purpose for writing the books of the "African Liberation S.M.A.T. Project" was to motivate the Creative

Energies, Minds, and Bodies of African people to go from an inert state of Theory and Speculation to an Active creative state of Development, Creation, and Productivity for the survival and well-being of African people everywhere. It is the goal of African Creation Energy's "African Liberation S.M.A.T. project" to free the minds, energies, and bodies of African people from mental captivity and physical reliance and dependence on inventions and technologies that were not developed or created by, of, and for African people.

In 2011, at the age of 30, after writing the books of the "African Liberation S.M.A.T. Project", Osiadan found it necessary to provide evidence of the **African Creation Energy** Philosophy in Action and Application by building structures, and thus embarked upon the project of building a Pyramid and authoring a text entitled "*ARCH I TET: How to Build A Pyramid*" as part of his **30 year "Djed Festival"** of renewal for all eyes to see.

In the summer of 2012, the book entitled "*9 E.T.H.E.R. R.E. Engineering*" was published which was dedicated to having readers better comprehend what they call their "**Spirit** and **Soul**" by studying various aspects of **Energy** including Electricity, Thermodynamics, Hydrodynamics, Electromagnetic Radiation, and Resonant Energy, and demonstrating operative use and practical applications of these various types of Energy as a form of "**Spirituality**".

In the Winter of 2012, the book "***Khnum-Ptah to Computer: The African Initialization of Computer Science***" was published as a way to provide motivation and inspiration for Africans and people of African descent to take part in the creation and development of software, computer programming, and computer-based technologies now and on into the future, and to also show the relationships between technological concepts like Computer Science, Robotics, Virtual Reality, and Transhumanism, to various traditional African cultural and spiritual concepts like animated Statues, and Death, and Resurrection.

At the Ascension age of 33 years old, Osiadan released the book "***H.E.R.U.-copters: African Aeronautical Ascension***" on 7-7-14 to show readers how the **African Creation Energy** philosophy can be used to "**go to Heaven**" by providing motivation and inspiration to Africans and people of African descent to study and utilize the scientific fields of Aviation, Aeronautics, and Aerospace Engineering.

And, in the year 2020, a year with symbolic significance to the concept of 20/20 visual acuity and one year before Osiadan's 40th birthday, the book "***SPEC-RA-SCOPE: The Symbolism and Significance of Spectroscopy to Science and Survival***" was released to present information on what can be viewed as the most important Science needed for survival. The plan is that the 9th and final **African Creation Energy** book will be released in the year 2021 when Osiadan has reached the age of 40 years old because it is the ancient tradition that at the age of 40 you are eligible to sit among the council of elders and be a "prophet". The plan is that the 9th and final **African Creation Energy** book will be an updated compilation of the previous books bound and packaged as a **Holy Tablet on African Science**.

Just like beliefs motivate and determine actions, the religion and spirituality of a people motivates and determines their level of science, math, technology, and creativity. The 3 great realities that a true spiritual system, doctrine, or way of life, must provide people in order for that system to be complete, are:
1. Hierarchy of Needs (food, clothing, shelter, education, entertainment, social interaction)
2. Morals and Ethics
3. Answers to Existential Questions

Science, Math, and Technology are the keys to providing people with methods to obtain all 3 of these "great realities". Thus, a true spiritual system, doctrine, or way of life must include science, math, and technology in order to be complete. One of the purposes of the "**African Creation Energy**" books is to show how Ancient and Traditional African culture and spiritual systems synthesized and combined Spirituality with Science, Math, and Technology. It is the desire of **African Creation Energy** to have people of African descent return to their traditional Scientific and Mathematic Spiritual systems. One of the most prolific examples of a Scientific and Mathematic Spiritual system is the spiritual system of the Ancient Africans in Egypt, which is why the **African Creation Energy** books have utilized Ancient Egyptian symbolism and terminology to teach Scientific, Mathematic, and Technological concepts. However, Ancient Egypt was a culture, society, and empire that spanned over 3000 years, had various and multiple groups of people as dynastic rulers, and over time contained practices, teachings, and beliefs which were positive and beneficial as well as negative and destructive. Let it be known, that the utilization of Ancient Egyptian symbolism and terminology to teach Scientific, Mathematic, and Technological concepts from an African perspective in the **African Creation Energy** books does not mean we are promoting everything Ancient and everything Egyptian as good. We recognize that second to the spiritual systems of Ancient Egypt, the religion of **Islam** is a spiritual system utilized by African people which has significantly impacted and influenced Scientific, Mathematic, and Technological development in the world; we also recognize that both of these spiritual systems have had their negative shortcomings as well. With the recognition that spiritual systems are created by human beings, for human beings, just like technology, then we must also recognize when that technology, or spiritual system, has become outdated, corrupted, and started to malfunction. When a Spiritual System, or technology, becomes outdated and starts to malfunction, it is the responsibility of the creators of that technology to **Upgrade** and **Update** the technology by using reason and discernment to select the best working parts of the technology, and throw away the corrupted malfunctioning parts of the technology. Likewise, a similar upgrade and update is needed at this point in time for the spiritual systems of African people worldwide. It is the goal of **African Creation Energy** to provide the upgrade and update to the African Scientific Spiritual System.

One of the more contemporary systems that people are familiar with, which combined spirituality and science, is called **Alchemy**, coming from the Arabic word *Al-Khemi*, meaning "**The Black**". Alchemical principles have been used in the development of the "African Creation Energy" books as a way to bring about the desired change and show how science and spirituality can coexist. The chosen term "**African Creation Energy**" has hidden Alchemical meaning. The word "**Africa**" comes from the Afro-asiatic word "**Afar**" meaning "**dust**" which represents the "**Earth**". The word "Africa" is also related to the Afro-Asiatic trilateral root F-R-Q, as in the word Furqan, meaning *"the criteria for discernment or separation"*, or the name Farooq meaning *"one who knows truth from falsehood"*. The etymology of the word "**Creation**" comes from the word "**Crescent**" which represents the "**Moon**", and the word "**Energy**" represents the "**Sun**", which is a primary source of energy to our planet. Therefore, one of the hidden meanings in the coined term "**African Creation Energy**" represents the "**Earth, Moon, and Sun**" cosmic forces also known as "**Ptah, Aah, and Re**" or "**Re Ah Ptah**" or "**Space, Matter, and Time**" in African Cosmology. Moreover, the abbreviation of "**African Creation Energy**" is **A.C.E.** which spells the word "Ace" which has etymological meanings of "a unit, whole, one, first, one who excels, and Primary" and indeed African Creation Energy represents the excellent Primary and **Original Creative Forces** in Nature.

The letters used to abbreviate "African Creation Energy", A.C.E., not only spell the English word "Ace" meaning "**First**, **Primary**, or **Original**", but also represent the 3 fundamental geometric shapes of the **Triangle** represented by the letter "**A**", the **Circle** represented by the letter "**C**", and the **Square** represented by the letter "**E**", which when combined form the Ancient **Alchemical symbol** of the "**Squared Circle**". All of the books authored by Osiadan Borebore Oboadee and **African Creation Energy** have been released on specific strategic dates, and each of the previously released books represents a different element necessary for transformation in Ancient African Alchemy. The Alchemical classical elements of Air, Water, Fire, and Earth are represented by the African Creation Energy books entitled *"The SCIENCE of Sciences, and The SCIENCE in Sciences"*, "Supreme *Mathematic African Ma'at Magic*", "*P.T.A.H. Technology*", and "*ARCH I TECT*" respectively. The quintessential element, or 5th element, of Ether or Spirit, is represented by the African Creation Energy book entitled "*9 E.T.H.E.R. R.E. Engineering*". The African Creation Energy book entitled "*Khnum-Ptah to Computer*" represents death, resurrection, and after-life, the book "*H.E.R.U.-copters*" represents "ascension into heaven", and the book "*SPEC-RA-SCOPE*" represents "divine perspective gained after ascension." These **8 A.C.E.** books provide a foundation for a scientific-based African spiritual system which can in turn be utilized to create a reality for your liberation. The "Science of Creating" is the information that is most needed by people of African descent right now and on into the future.

The Symbolism & Significance of Spectroscopy to Science & Survival

If a Technology or ANY creation or invention is being used by a group of people, and that Group of People is dependent on that technology for Survival and Well Being, but that Group of people is not in control of the Creation or Production of that Technology, then that Group of People are Literally **SLAVES** to the Creators and the Producers of the Technology. As Africans and people of African descent look to the Future and we see the Rapid Advances of Technology, but we do not see ourselves as the Developers, Creators, Inventors, Designers, or Producers of the Technology, then a sense of Hopelessness, despair, desperation, anger, helplessness, Powerlessness, and feelings of oppression arise. **African Creation Energy** has shown that we as African people have always been **Creators**, **Fashioners**, and **Makers** of Technology and we do indeed have a place as the Developers, Inventors, and Designers of all the highly advanced technologies which will shape the Future. In the near Future, the ability to be **conscious**, or **aware**, or **knowledgeable** about something will become trivial, and thus the next level of consciousness, will become **ACTIVENESS** and having the ability to use and put all of the knowledge and information which one is conscious and aware of, into practical application. No longer will "Knowing" be a big deal because almost everyone will "KNOW", the next level will be "**how can what is known be used and applied**". Knowledge is a natural resource. It is the practical application and utilization of all natural resources, including knowledge, which determines economic prosperity. Following the plethora of information presented by the many great African Scholars (who have affectionately been labeled "**MASTER TEACHERS**") who have came to improve the conditions of African people; it is the goal of **African Creation Energy** to be the catalyst in the synthesis, unification, and practical application of the information presented by the great Master Teachers. Thus, it is the aspiration of "African Creation Energy" to be, and breed, "**Master Technicians**" who TEaCH through Action and Application.

African Creation Energy (A.C.E.) is dedicated to **African Centered Education** (A.C.E.) and one of the goals of African Creation Energy is to introduce the utilization of African terminology and symbolism in Science, Technology, Engineering, and Math (S.T.E.M.) education. African Creation Energy is the **ROOT of S.T.E.M.** (Science, Technology, Engineering, and Math).

Scientist, Mathematician, Electrical Engineer, Architect, Computer Programmer, Roboticist, Aviator, and Spectroscopist Osiadan is a *"Jack of all trades and a Master of Nun"* (Nun is the Ancient African Egyptian name for the Original Creative Forces in Nature). "**Africa**" (F-R-Q) is wherever we are in our **right mind**; and this includes on other continents or other planets. It is a fact that our sun and planet Earth will not always be in a habitable condition, and therefore the science of space exploration is a science necessary for the Survival and wellbeing of African people in the future. "**Afro-futurology**" is an "African-centered" scientific and mathematic-based "prophecy" methodology to *"foretell the future"*. In the future, as Africans began to explore the cosmos, we will go from being "Afro-Centric" to being "**Astro-Centric**", and **African Creation Energy** is the Escalator, the Elevator, and the Aviator which raises you up and takes you higher.

The Symbolism & Significance of Spectroscopy to Science & Survival

African Creation Energy Books

HERU-COPTERS:
African Aeronautical Ascension
Release Date: 07-07-14

SPEC-RA-SCOPE:
The Symbolism & Significance of Spectroscopy to Science & Survival
Release Date: 12-20-20

ARCH I TECT:
How to Build A Pyramid
Release Date: 11-11-11

9 E.T.H.E.R. R.E. Engineering
Release Date: 06-26-12

KHNUM-PTAH To COMPUTER:
The African Initialization of Computer Science
Release Date: 12-12-12

The SCIENCE of Sciences and The SCIENCE in Sciences
Release Date: 10-10-10

9^{9^9} **Supreme Mathematic, African Ma'at Magic**
Release Date: 09-09-09

P.T.A.H. Technology:
Engineering Applications of African Sciences
Release Date: 05-04-10

9. Appendix

REFERENCES & RESOURCES

1. "9^{9^9} Supreme Mathematics African Ma'at Magic" By African Creation Energy
2. "9 E.T.H.E.R. R.E. Engineering" By African Creation Energy
3. "African Philosophy: The Pharaonic Period: 2780-330 BC" by Dr. Théophile Obenga
4. "Ambient mass spectrometry using desorption electrospray ionization (DESI): instrumentation, mechanisms and applications in forensics, chemistry, and biology" - Zoltan Takats, Justin M. Wiseman and R. Graham Cooks, Purdue University, 2005
5. "ARCH I TECT: How to Build A Pyramid" By African Creation Energy
6. "Conspiracy theories: Evolved functions and psychological mechanisms" by Van Prooijen and Van Vugt
7. "Egyptian Hieroglyphic Dictionary Vol. 1" by E.A. Wallis Budge
8. "HERU-COPTERS: African Aeronautical Ascension" By African Creation Energy
9. "The Isis Papers" by Dr. Frances Cress Welsing
10. "KHNUM-PTAH To COMPUTER: The African Initialization of Computer Science" By African Creation Energy
11. "The Misinformation Age: How False Beliefs Spread" by Cailin O'Connor and James Owen Weatherall
12. "P.T.A.H. Technology: Engineering Applications of African Sciences" By African Creation Energy
13. Public Lab contributors (https://publiclab.org/wiki/foldable-spec)
14. "The SCIENCE of Sciences and the SCIENCE in Sciences" By African Creation Energy
15. Spectral Work Bench (https://spectralworkbench.org)
16. "Stolen Legacy" by George G.M. James

Printed in Great Britain
by Amazon